Ricardo Gattass •
Juliana G.M. Soares •
Bruss Lima

The Pulvinar Thalamic Nucleus of Non-Human Primates: Architectonic and Functional Subdivisions

Springer

Ricardo Gattass
Laboratory of Cognitive Physiology
Institute of Biophysics Carlos Chagas Filho
Federal University of Rio de Janeiro
Rio de Janeiro, Brazil

Juliana G.M. Soares
Laboratory of Cognitive Physiology
Institute of Biophysics Carlos Chagas Filho
Federal University of Rio de Janeiro
Rio de Janeiro, Brazil

Bruss Lima
Laboratory of Cognitive Physiology
Institute of Biophysics Carlos Chagas Filho
Federal University of Rio de Janeiro
Rio de Janeiro, Brazil

ISSN 0301-5556 ISSN 2192-7065 (electronic)
Advances in Anatomy, Embryology and Cell Biology
ISBN 978-3-319-70045-8 ISBN 978-3-319-70046-5 (eBook)
https://doi.org/10.1007/978-3-319-70046-5

Library of Congress Control Number: 2017957219

Cover illustration: deblik, Berlin

Printed on acid-free paper

This Springer imprint is published by Springer Nature
The registered company is Springer International Publishing AG
The registered company address is: Gewerbestrasse 11, 6330 Cham, Switzerland

Advances in Anatomy, Embryology and Cell Biology publishes critical reviews and state-of-the-art surveys on all aspects of anatomy and of developmental, cellular and molecular biology, with a special emphasis on biomedical and translational topics.

The series publishes volumes in two different formats:

- Contributed volumes, each collecting 5 to 15 focused reviews written by leading experts
- Single-authored or multi-authored monographs, providing a comprehensive overview of their topic of research

Manuscripts should be addressed to
Co-ordinating Editor

225

Advances in Anatomy, Embryology and Cell Biology

More information about this series at
http://www.springer.com/series/102

Abstract

Traditionally, the thalamus is considered a simple relay station controlling the transmission of sensory information from the periphery to the cortex. This may be partly true for the so called "first-order" nuclei, such as the lateral geniculate nucleus (LGN), which receive signals from retinal ganglion cells and transmit them to the primary visual cortex. However, recent studies have challenged this simplistic view of thalamic function, particularly concerning its "higher-order" nuclei, such as the pulvinar, which receive strong driver input from the cortex. In addition to receiving cortical afferents, the pulvinar also projects back to several cortical areas, suggesting that it may serve a role in indirect cortical-cortical communication via the thalamus (in parallel to the more direct cortical-cortical pathways). Visuotopic organization is an important feature of several of the cortical visual areas that are reciprocally interconnected with the pulvinar. Accordingly, the pulvinar also exhibits visuotopic organization, probably reflecting its pattern of connectivity with the cortex and the superior colliculus (SC). However, while individual cortical areas usually have a single map of the visual world, multiple visuotopic maps can be present in the pulvinar. How do these maps relate to the pattern of pulvinar-cortical connectivity and to the anatomical subdivisions of the pulvinar? Within this framework, it is paramount that we understand the functional, anatomical, and connectional organization of the pulvinar in order to evaluate its role in cortical information processing. To this aim, we use an evolutionary approach where we compare how pulvinar organization evolved among New World and Old World monkeys. We argue that the chemoarchitectural organization of the pulvinar is highly conserved along primate evolution. On the other hand, pulvinar functional organization, as evidenced by the number and arrangement of topographically organized visual maps, is markedly variable across species. We are thereby challenged by the fact that diverse functional maps, which probably evolved because of distinct selective and behavioral pressures, are able to coexist within a common chemoarchitectural scaffold. We also discuss the comparative cyto- and myeloarchitecture of the pulvinar across different primate species and argue that both descriptions reside somewhat in between the rigid chemoarchitectural scaffold on the one hand and the flexible functional layout of the pulvinar on the other. As such, the cyto- and myeloarchitectures of the pulvinar only partially match the pattern of pulvinar-cortical connectivity, while electrophysiological and connectivity data are highly consonant with one other. To support our claims, we review electrophysiological, anatomical, and chemo-, cyto-, and myeloarchitectural data of the pulvinar in different primate species, with special focus on the New World capuchin monkey and the Old World macaque monkey. We finish by discussing the potential role of the pulvinar as a functional hub capable of organizing brain activity across different brain systems and across distant regions. The roles of the pulvinar include the control of eye movements, the selection of salient stimuli, and the modulation of attention, suggesting that the pulvinar may act to gate incoming information to the neocortex. Indeed, early studies appointed the pulvinar as a link between the retinotectal system and the retino-geniculate-cortical system. Later studies have highlighted the role of the pulvinar as an integration center. The emerging view is that the pulvinar may be critical to several integrative aspects of cognition, such as modulating cortical arousal and the effective allocation of attention.

Contents

Chapter 1
Introduction

The optic tracts terminate in a structure denominated as the *thalami nervorum opticorum*, or the "optic nerve chamber." This large gray mass of nervous tissue located at the base of the brain is now commonly designated as the thalamus. The thalamic nuclei are assigned into the anterior, medial, lateral, ventral, and posterior nuclear groups, in addition to the midline nuclei. The pulvinar comprises the largest portion of the posterior group, which extends from the habenular complex to the caudal end of the thalamus (Walker 1938). In primates, the pulvinar is also the largest thalamic nucleus. During evolution, parts of the pulvinar seem to have evolved from a telencephalic anlage. The development and differentiation of the pulvinar paralleled a major neocortical enlargement encompassing the temporal, parietal, and occipital regions. Accordingly, the pulvinar is heavily interconnected with several cortical areas (Siqueira 1971; Harting et al. 1972; Rakic 1974).

The pulvinar can be subdivided into well-delimitated regions based on chemoarchitectural, cytoarchitectural, myeloarchitectural, connectivity, and electrophysiological criteria (see Table 1.1). However, the subdivisions that emerge from these various techniques do not always match one another. The classical work by Walker (1938) used cytoarchitectural criteria to subdivide the pulvinar of the macaque monkey into medial, lateral, and inferior portions. However, this classical delimitation does not correspond to those proposed later by other researchers based on connectivity (e.g., Lin and Kaas 1979; Ungerleider et al. 1984; Soares et al. 2001), electrophysiological topographic mapping (Allman et al. 1972; Gattass et al. 1978a; Bender 1981), or chemoarchitecture (Cusick et al. 1993; Steele and Weller 1993; Gutierrez et al. 1995; Stepniewska and Kaas 1997; Adams et al. 2000).

Subdivisions of the pulvinar based on its chemoarchitectural features are the most consistently preserved across species of New and Old World monkeys (i.e., compare the subdivisions revealed using immunocytochemistry across macaque, capuchin, squirrel, and owl monkeys in Table 1.1). Immunocytochemical staining for calbindin, for example, is one of the methods generally employed to reveal the chemoarchitecture of the primate brain. In the neocortex, calbindin is observed

R. Gattass et al., *The Pulvinar Thalamic Nucleus of Non-Human Primates: Architectonic and Functional Subdivisions*, Advances in Anatomy, Embryology and Cell Biology 225, https://doi.org/10.1007/978-3-319-70046-5_1

Table 1.1 Pulvinar partitioning based on different methods

Species	Method	Pulvinar subdivisions									Source
Rhesus monkey	Cytoarchitecture	PM	PI					PL			(1)
	Electrophysiology					PI		PL			(2)
	Connectivity		P3			P1		P2			(3)
	Imunocytochemistry	PM	PI_P	PI_M	PI_C	PI_L			PI_{LS}	PL	(4, 5, 6)
		PM	PI_P	PI_M	PI_{CM}	PI_{CL}		PL			(7)
		PM	PI_P	PI_M	PI_{CM}	PI_{CL}		PL_{VM}	PL_{VL}		(8)
Capuchin monkey	Electrophysiology					P_{VL}				P_μ	(9)
	Connectivity			P3		P1				P4	(10)
	Imunocytochemistry	PM		PI_P	PI_M	PI_C		PI_L	PI_{LS}	PL	
Squirrel monkey	Cytochemistry	PM	PL_M	PI_P	PI_M	PI_C				PL_L	(11)
	Immunocytochemistry	PM		PI_P	PI_M	PI_C		PI_L		PL	(12)
		PM		PI_P	PI_M	PI_C	PI_L	PI_{LS}		PL	(13)
Owl monkey	Electrophysiology			PI							(14)
	Cytoarchitecture and connectivity			IP_P	IP_M	IP_C					(15)
	Immunocytochemistry	PM		PI_P	PI_M	PI_{CM}	PI_{CL}	PL			(16)

(1) Walker (1938); (2) Bender (1981); (3) Ungerleider et al. (1984); (4) Cusick et al. (1993); (5) Gutierrez et al. (1995); (6) Gray et al. (1999); (7) Stepniewska and Kaas (1997); (8) Adam et al. (2000); (9) Gattass et al. (1978a); (10) Soares et al. (2001); (11) Steele and Weller (1993); (12) Cusick et al. (1993); (13) Gray et al. (1999); (14) Allman et al. (1972); (15) Lin and Kaas (1979); (16) Stepniewska and Kaas (1997)

primarily in a subset of inhibitory GABAergic interneurons, but it can also be present in non-GABAergic pyramidal cells (Baimbridge et al. 1992; Andersen et al. 1993; DeFelipe 1997; Hof et al. 1999). Calcium-binding proteins are thought to buffer intracellular calcium levels, which are important for signaling within the cell and for modulating its activity (Andersen et al. 1993). Although the functional role of calbindin-immunoreactive neurons in the thalamus is yet unknown, they may also be important in modulating neocortical activity (Jones and Hendry 1989; Hashikawa et al. 1991), as suggested by the pattern of pulvinar-cortical projecting neurons rich in calbindin. It is reasonable to speculate that the occurrence and distribution of calcium-binding proteins in the pulvinar, such as calbindin and parvalbumin, have been preserved along evolution. Therefore, they have proven to be valuable tools capable of probing the basic pulvinar scaffold across primate species. Along this review, we will provide an overview of the available data regarding the various subdivisions of the pulvinar that have been proposed based on architectural criteria such as the distribution of molecular markers, neuronal morphology, and fiber layout.

Our ultimate goal is to understand the functional organization of the pulvinar. Descriptions on its architectural layout are valuable, but it must be corroborated with functional data capable of revealing the response characteristics of pulvinar neurons. Particularly relevant are its sensory response properties, how they participate in key cognitive functions (such as arousal systems or the allocation of attention), and how the pulvinar relates to the rest of the brain, potentially serving as a hub for the integration of neuronal processing. To this aim, we will review electrophysiological data relating to the visual topographic organization of the pulvinar, which suggests that it contains at least four representations of the contralateral visual field, each placing different emphasis on central vs. peripheral visual field representation. Subsequently, we will present data on pulvinar-cortical connectivity and how it relates to the maps obtained using electrophysiological recordings. We will focus particularly on the connectivity of the pulvinar with the topographically organized visual areas V1, V2, MT, and V4, for which we have extensive results. Based on tracing studies, we find that data on the electrophysiological mapping of the pulvinar agrees with the pattern of pulvinar-cortical connectivity. However, it is challenging to infer pulvinar visual topography based exclusively on the connectivity pattern due to the widespread overlap of these projections.

Subsequently, we illustrate the diversity of neuronal response properties as measured by electrophysiological recordings in the pulvinar. We observe from concentric *on-off*, retina-like receptive fields, all the way to orientation and direction-selective neurons, including hypercomplex-type cells. We review data on GABA inactivation of the pulvinar and how the latter affects neuronal selectivity in area V2. GABAergic inactivation of the pulvinar also affects overall activity levels in V2. Together with other recently published results, we argue that the pulvinar participates in a general arousal system and has thereby an important role in activating the cortex.

Finally, we address the possible role of the pulvinar in selective visual attention. Due to its widespread connectivity pattern with the cortex, along with the abundant projections it receives from the SC, the pulvinar is ideally placed to coordinate large-scale attentional networks. Notably, the pulvinar contains visual maps that are reciprocally and topographically connected with several visual cortical areas. This characteristic enables the pulvinar to function as a coordinating hub, capable of integrating the processing of neuronal information across different brain regions, specifically for those stimuli being presented inside the attentional focus. We still lack a clear neuronal mechanism capable of implementing such a proposal. We discuss the potential role of synchronous oscillations as a unifying mechanism allowing long-range inter-areal neuronal communication.

Chapter 2
Cytoarchitecture and Myeloarchitecture of the Pulvinar

Walker (1938) subdivided the pulvinar of the macaque monkey into three portions based on topography and cytoarchitecture. The *nucleus pulvinaris medialis* (PM) is the medial and larger portion of the pulvinar, with compactly arranged polygonal cells and only a few transverse fibers. The *nucleus pulvinaris lateralis* (PL) comprises the lateral portion of the pulvinar and blends with the PM without a clear demarcation border. PL cells are rather small, easily stained, polygonal in shape, and are segregated in clumps by the many horizontally crossing fibers. The *nucleus pulvinaris inferior* (PI) lies in the ventral posterior portion of the pulvinar, between the medial (MGN) and the lateral (LGN) geniculate bodies. Laterally, it is bordered by the PL and dorsally by the brachium of the (SC). It is composed of compactly arranged small, dark, polygonal-shaped cells. Olszewski (1952) extended the anterior limits of the pulvinar and added a subdivision to this nucleus, named *pulvinar oralis* (PO). PO appears between the *nucleus centrum medianum* (CM) and the *nucleus ventralis posterior lateralis* (VPL). The cells are small and lightly stained and exhibit irregular density being, in general, less cellular than the other portions of the pulvinar.

In New World marmoset (*Callithrix jacchus*), squirrel (*Saimiri sciureus*), and capuchin (*Sapajus apella*, formerly *Cebus apella*) monkeys, we observe a similar subdivision of the pulvinar (Eidelberg and Saldias 1960; Mathers 1972; Spatz and Erdmann 1974; Soares et al. 2001). Figure 2.1 shows the cytoarchitecture of the pulvinar of the capuchin monkey in four different coronal planes, spaced 0.5 mm apart.

Mathers (1972) studied the ultrastructure of the pulvinar of the squirrel monkey and showed two populations of neurons that were quite similar to those found in other thalamic nuclei. The first neuron type is the thalamocortical relay cell (TRC), averaging 26 μm in diameter and exhibiting radial dendritic pattern and thin dendritic appendages. The second type is the Golgi type II neuron, about 16 μm in diameter, with more complex dendritic appendages and an axon ramifying in the vicinity of the soma (Fig. 2.2). The TCR-Golgi type II ratio is about 7:3, in both PL

R. Gattass et al., *The Pulvinar Thalamic Nucleus of Non-Human Primates: Architectonic and Functional Subdivisions*, Advances in Anatomy, Embryology and Cell Biology 225, https://doi.org/10.1007/978-3-319-70046-5_2

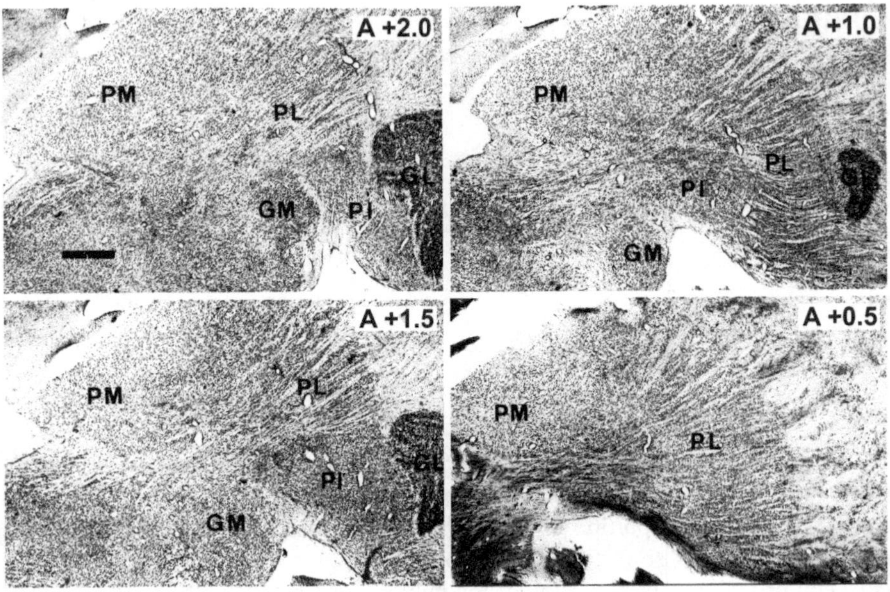

Fig. 2.1 Cytoarchitectonic subdivisions of the pulvinar according to Olszewski (1952). The cytoarchitectonic subdivisions are overlaid onto Nissl stained coronal sections of the capuchin monkey brain (right hemisphere), following the rostral (top left)-to-caudal (bottom right) extent of the pulvinar. The sections are spaced 0.5 mm apart, and they do not reach the caudal extent of the pulvinar. Abbreviations: *GM* medial geniculate nucleus, *GL* lateral geniculate nucleus, *PL* lateral pulvinar, *PM* medial pulvinar, *PI* inferior pulvinar. Scale bar = 1 mm. (Modified from Soares et al. 2001)

and PI, and 9:2 in PM. The most commonly found terminal is small with round vesicles (RS terminal). They are distributed throughout the neuropil, usually making contact with dendritic profiles. There are a significant number of axo-axonal synapses where a large terminal with round vesicles (RL terminal) is presynaptic to a terminal with flattened vesicles (F terminal). However, the large majority of the synapses are the usual axo-dendritic synapses. RL terminals nearly always make synaptic contact with more than one structure forming the so-called glomeruli.

Ogren and Hendrickson (1979) studied the structural organization of PI and PL of the macaque monkey pulvinar and described two neuronal types. The projection neurons (PN) vary in cell body (15–40 μm) and dendritic tree (150–600 μm) diameters but bear the same variety of dendritic appendages, namely, spine-like, hairlike, and knot-like. The local circuit neurons (LCN) have smaller cell body diameters (10–20 μm) but can have very large dendritic field diameters (150–600 μm). They are best distinguished from PN by their elaborate dendritic appendages, which have been identified as presynaptic dendrites under electron microscopy (EM). They also described four types of synaptic terminals. RS and RL terminals both contain round synaptic vesicles and make asymmetric synaptic contacts. RL contact larger-caliber dendrites and frequently form synaptic

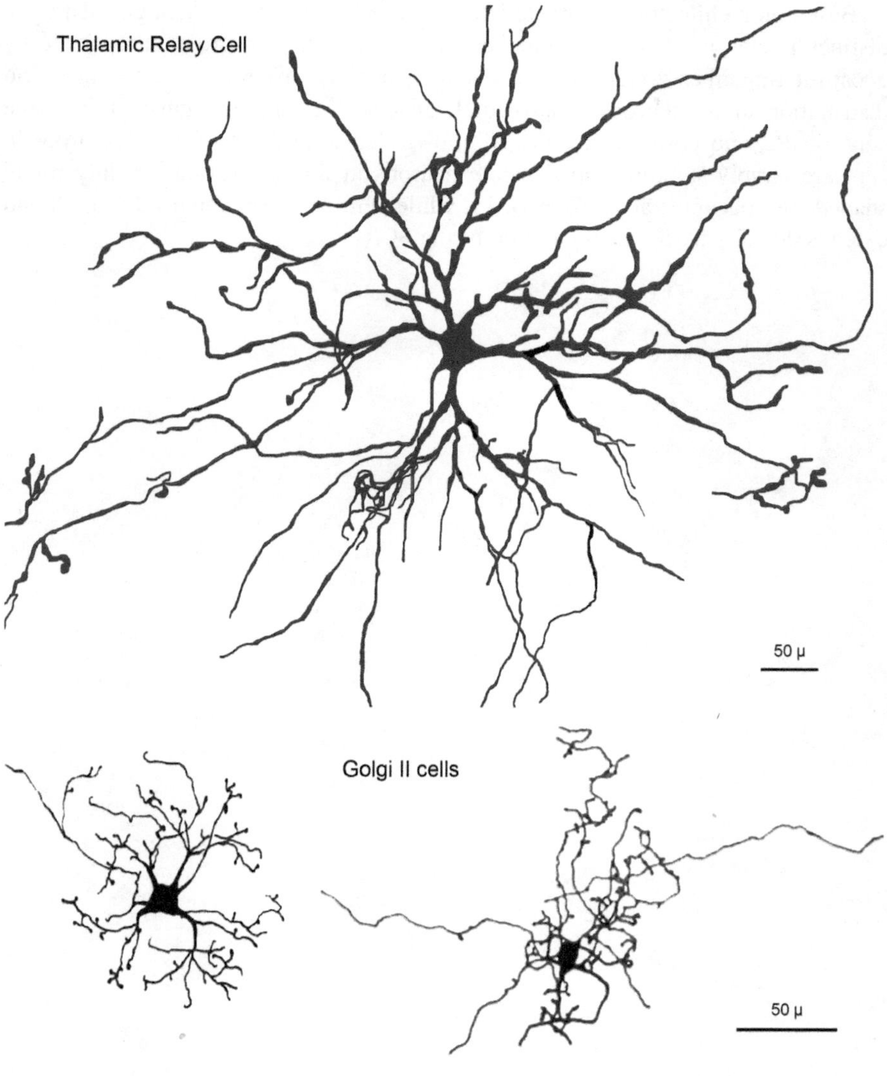

Fig. 2.2 Reconstruction of one thalamocortical relay cell (TRC) and two Golgi Type II cells from the PI (inferior pulvinar) and PL (lateral pulvinar) subdivisions of the pulvinar. The neurons were impregnated using the Golgi-Kopsch method. Top: TCR cell with the radial dendritic arbor, and some of the filiform spines that typify this neuronal type. Bottom: Drawings of two Golgi type II neurons with their more complex dendritic appendages. Scale bars = 50 μm. (Based on data from Mathers 1972)

complexes with presynaptic dendrites of LCN, while RS contact fine-caliber dendrites and only rarely take part in synaptic complexes. F terminals and P boutons both contain flat and pleomorphic vesicles and make symmetric synaptic contacts. The quantitative distribution of each type is very similar in both subdivisions, averaging 85% for RS, 5% for RL, 0.3% for F, 8% for P, and 2% unidentified.

Based on architectonic characteristics, Lin and Kaas (1979) distinguished three distinct nuclei in PI of owl monkeys that are separated from each other by encapsulating fiber zones and are distinguished by differences in the size and distribution of its neurons. The medial nucleus, IPm, is distinguished from the central, IPc, and posterior nucleus, IPp, by a denser packing of cells. Neurons in IPm are largely spindle shaped, while neurons in IPc and IPp are mainly round shaped. IPc occupies about 70% of PI, while IPm occupies about 20% of PI and extends dorsally across the brachium of the SC.

Chapter 3
Chemoarchitecture of the Pulvinar

Cytochemical and immunocytochemical methods reveal details of the pulvinar architecture that are not apparent from Nissl and myelin staining. However, the results of these techniques have been interpreted in different ways by different investigators, each adopting different sets of nomenclature for the various pulvinar subdivisions.

3.1 Chemoarchitecture in the Macaque Monkey

In the macaque, Lysakowski et al. (1986) described a dense acetylcholinesterase (AChE) reactivity in regions of PI and PL that coincide with the projections zones of the SC described by Benevento and Standage (1983). Studies of the chemoarchitecture of the pulvinar in macaque monkeys using histochemical reactivity for the cytochrome oxidase (CO) and AChE enzymes and immunostaining for calbindin (Cb), parvalbumin (Pv), SMI-32, Cat-301, and Wisteria floribunda agglutinin (WFA) (Cusick et al. 1993; Gutierrez et al. 1995; Gray et al. 1999) revealed five subdivisions of PI, which include all of the traditional PI but which also encompass parts of PL and PM. Two of these subdivisions match those described by Lin and Kaas (1979), namely, the posterior (PI_P) and medial (PI_M) subdivisions. However, the central or "classic" portion was subdivided into central (PI_C) and lateral (PI_L) subdivisions. PI_M, a calbindin-poor zone located between PI_P and PI_C (regions with intense staining for calbindin) presented elevated CO and AChE activity, elevated SMI-32 and parvalbumin immunostaining, and dense patches of WFA and Cat-301 immunostaining. The intense patches of AChE staining within PI_M closely resemble the dense AChE-positive bands that overlap zones of tectal input, as described by Lysakowski et al. (1986). PI_C was characterized by a moderate AChE and a light CO activity, lightly stained for parvalbumin, very lightly stained for WFA, and with only few SMI-32 labeled neurons. PI_P, a small region located more rostrally, exhibited light AChE activity, was very lightly stained for WFA and was unstained for SMI-32.

R. Gattass et al., *The Pulvinar Thalamic Nucleus of Non-Human Primates: Architectonic and Functional Subdivisions*, Advances in Anatomy, Embryology and Cell Biology 225, https://doi.org/10.1007/978-3-319-70046-5_3

In addition, PI_P, in the sections stained for CO or parvalbumin, was distinctly paler than the adjacent PI_M. PI_L adjoins the LGN and lightly stains for calbindin, parvalbumin, and CO. Gutierrez et al. (1995) described yet in macaque monkeys a most lateral portion, termed the shell of PI_L (PI_{LS}), which appeared to contain many dendrites intensely stained for calbindin, in addition to numerous WFA-stained neurons, when compared to more medially adjacent portions of PI_L. The medial border of PI_{LS} is often characterized by large calbindin stained cells.

Stepniewska and Kaas (1997), based on the same techniques used by Cusick and collaborators (Cusick et al. 1993; Gutierrez et al. 1995; Gray et al. 1999), proposed a new subdivision for PI where PI_C, which does not include part of the cytoarchitectonic PL subregion, was subdivided in PI_{CL} and PI_{CM} (Table 1.1).

Figure 3.1 shows enlarged micrographs of calbindin immunoreactive cells in the pulvinar of the macaque monkey (Adams et al. 2000). PI_P and PI_{CM} were the

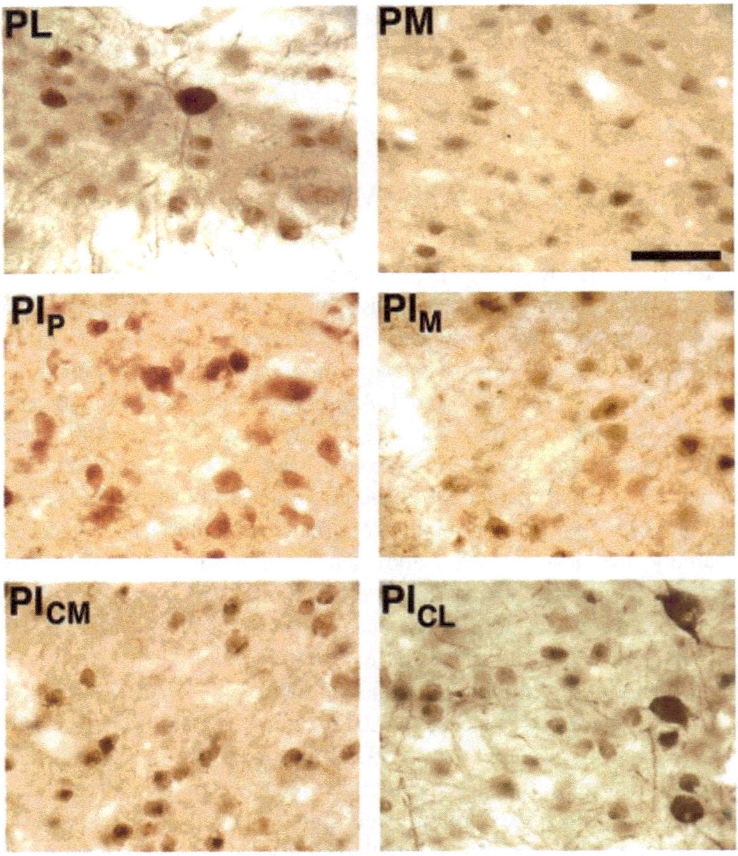

Fig. 3.1 Chemoarchitecture of the macaque monkey pulvinar. High magnification photomicrographs of calbindin-positive cells in the different subdivisions of the pulvinar. Each photomicrographs shows calbindin immunoreactive cells in PL, PM, PI_P, PI_M, PI_{CM}, and PI_{CL}. See text for details. Scale bar = 200 μm [modified from Adams et al. (2000)]

regions showing the strongest calbindin immunoreactivity (note the darkly stained small cells and dense neuropil that are revealed by the reaction). The difference between the two regions was that the PI_{CM} zone contained a few large cells (not shown in figure), which were absent in PI_P. The PI_M subdivision, the calbindin-poor PI region located between PI_P and PI_{CM}, contained very little neuropil staining and only a few faintly labeled cells. The PI_{CL} zone showed moderate calbindin staining. Within the latter, there was a dense population of small calbindin-containing neurons and moderate neuropil staining, as well as a small number of large stained neurons. The small calbindin-containing neurons had a clear stained soma with a few visible dendrites, while the larger immunoreactive neurons had a darkly stained soma with radiating dendrites. Relative to PI_M, PI_{CL} had denser neuropil staining and more darkly stained calbindin immunoreactive neurons, but the staining was not as intense as in PI_P and PI_{CM}. The distinguishing feature of PI_{CL} was the numerous large cells that were scattered throughout this subdivision. The ventral portion of the cytoarchitectonic PL was similar to PI_{CL} in that both of these regions had a large number of small calbindin-containing neurons and a few scattered large cells. Although Cusick et al. (1993) and Gutierrez et al. (1995) considered this subdivision as part of PI_{CL} (PI_L in their terminology), PI_{CL} and PL could be distinguished by the presence of horizontally oriented fiber bundles in the latter. Dorsally in the pulvinar, cytoarchitectonic PM also contained many small calbindin-positive neurons and a few scattered large neurons (not shown in figure). However, the neuropil in PM was more intensely stained than in PL.

Gutierrez et al. (2000) described a region in the dorsal portion of PL, named PL_D, which stained dark and uniform for AChE and parvalbumin, but appeared pale with calbindin immunostaining. The ventromedial border between the neurochemical subdivision PL_D and the rest of the dorsal pulvinar, termed the medial pulvinar (PM), could be sharply defined. AChE and parvalbumin reactivities were weaker for PM compared to PL_D and displayed both lateral (PM_L) and medial (PM_M) histochemical divisions. PM_M contained a central "oval" region ($PM_{M\text{-}C}$) that stained darker for AChE and parvalbumin than the surrounding region.

3.2 Chemoarchitecture in New World Monkeys

Steele and Weller (1993), based on AChE activity, divided PL of the squirrel monkey into a lateral, darker staining subdivision, denominated PL_L, and a medial one, PL_M. Additionally, in accordance with Lin and Kaas (1979), they subdivided PI into a "classic" lateral region (PI_C), a "middle" region (PI_M), and a posterior region (PI_P). PI_M is located between PI_C and PI_P, with the former displaying a weak staining and the latter two subregions exhibiting a relatively strong staining for AChE. This AChE staining pattern for PI_M is opposite to the PI_M staining pattern described for the squirrel monkey by Gray et al. (1999) and Stepniewska and Kaas (1997). These authors, however, found in PI of the squirrel and owl monkeys a pattern similar to the one described for the macaque, with small variations. Squirrel

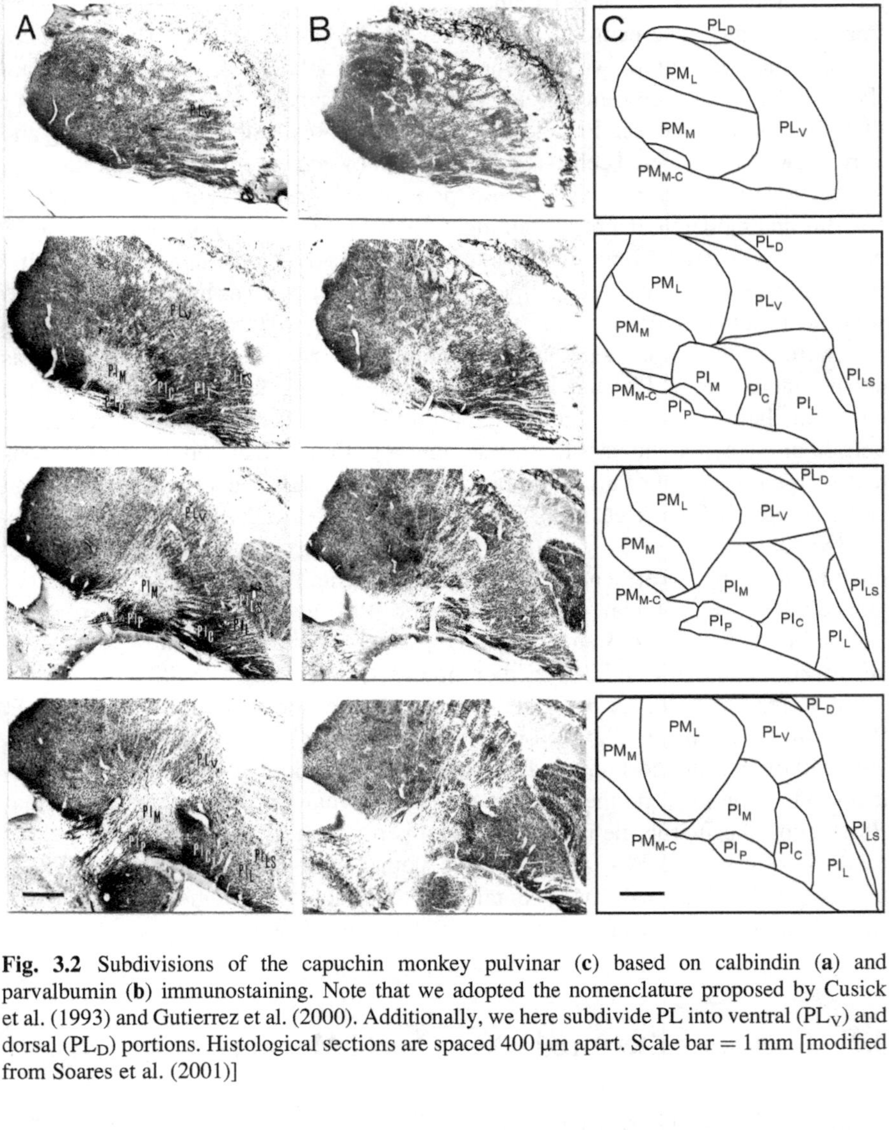

Fig. 3.2 Subdivisions of the capuchin monkey pulvinar (**c**) based on calbindin (**a**) and parvalbumin (**b**) immunostaining. Note that we adopted the nomenclature proposed by Cusick et al. (1993) and Gutierrez et al. (2000). Additionally, we here subdivide PL into ventral (PL_V) and dorsal (PL_D) portions. Histological sections are spaced 400 μm apart. Scale bar = 1 mm [modified from Soares et al. (2001)]

and macaque monkeys showed opposite staining patterns within PI_C and PI_M for the Cat-301 antibody.

In the pulvinar of the capuchin monkey, Soares et al. (2001) described a chemoarchitectonic pattern similar to the one described in the macaque monkey by Cusick and collaborators (Cusick et al. 1993; Gutierrez et al. 1995). The immunohistochemical staining for calbindin and parvalbumin in the pulvinar of the capuchin monkey is illustrated in Fig. 3.2. Based on immunohistochemical staining, mainly for calbindin, the border of PI was shifted dorsally, above the brachium of the SC. Additionally, PI was subdivided into five regions (PI_P, PI_M,

Fig. 3.3 Chemoarchitecture of the capuchin monkey pulvinar. Enlarged photomicrographs of adjacent coronal sections of the pulvinar (shown in Fig. 3.2) reacted for the calcium-binding proteins calbindin and parvalbumin [modified from Soares et al. (2001)]

PI_C, PI_L, and PI_{LS}). Figure 3.3 shows two enlarged sequential coronal sections of the capuchin monkey pulvinar stained for calbindin and parvalbumin. Under calbindin staining, the original inferior pulvinar can be further subdivided into PI_P, PI_M, and PI_C. Both the PI_P and PI_C zones display the heaviest calbindin immunoreactivity. The PI_M is almost devoid of calbindin immunoreactivity. On the other hand, the staining for parvalbumin shows a heavily labeled PI_M and a homogenous staining for PI_C, PI_L and PI_{LS}. PI_P is only lightly labeled by parvalbumin.

The capuchin pulvinar, specially the intermediate and posterior portions of PI and the lateral portion of PL, also react strongly for AChE. This pattern of AChE staining resembles that described in the macaque (Lysakowski et al. 1986) and squirrel (Steele and Weller 1993) monkeys, but it differed from the pattern described by Gray et al. (1999), where PI_M stained strongly for AChE.

Figure 3.4 shows enlarged photomicrographs of portions of PI stained for the SMI-32 antibody. SMI-32, a monoclonal antibody that recognizes a nonphosphorylated epitope on neurofilament proteins (Sternberger and Sternberger 1983), has been used to define regional patterns of cortical organization in the visual system (Hof and Morrison 1995). In the pulvinar as a whole, the immunocytochemistry for SMI-32 shows a light staining pattern. However, we find some large, heavily labeled neurons scattered throughout PM and PI but mainly throughout PL (Fig. 3.4b). These cells are similar to the large calbindin-positive neurons that are present in similar locations of the pulvinar but that are fewer in number. The medial portion of PI shows a darker staining as well as some moderately labeled medium-sized SMI-32 cells (Fig. 3.4a).

In spite of the differences in nomenclature that have been proposed by the various authors, a similar chemoarchitectonic pattern is revealed by calbindin reactions in all primate species studied so far. There is an agreement relative to the borders of PI_P and PI_M and of the darker adjacent region named PI_C (Gutierrez et al. 1995; Gray et al. 1999; Soares et al. 2001) or PI_{CM} (Stepniewska and Kaas 1997; Beck and Kaas 1998; Adams et al. 2000). In addition, all these authors

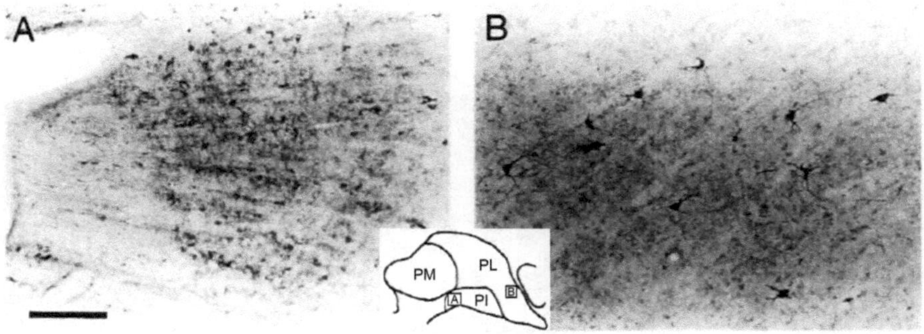

Fig. 3.4 Chemoarchitecture of the capuchin monkey pulvinar. Enlarged photomicrographs of a coronal section of the pulvinar (right hemisphere) reacted with the SMI-32 antibody. Note the very distinct reactivity patterns in the PI (i.e., PI_P) and the PL (i.e., PI_L) subdivisions (panels (**a**) and (**b**), respectively). Scale bar = 30 μm [modified from Soares et al. (2001)]

reinforce the idea that these subdivisions cross the limits of the brachium of the SC, occupying part of the adjacent PM and/or PL. The major controversy is related to the partitioning of the ventrolateral portion of the pulvinar. Cusick and colleagues (Cusick et al. 1993; Gutierrez et al. 1995; Gray et al. 1999), based on the similarity of calbindin staining patterns in the lateral portion of PI and in the ventral portion of PL and by the fact that the V1 projection zone extends dorsal to the brachium of the SC, consider this region as a single subdivision named PI_L. However, other authors (Stepniewska and Kaas 1997; Beck and Kaas 1998; Adams et al. 2000) prefer to maintain the original subdivisions proposed by Lin and Kaas (1979) and subdivide PI_C into PI_{CM} and PI_{CL}. Although these authors recognize that PI_{CM} extends above the brachium of the SC, they assume PI_{CL} to be restricted to the lateral portion of the traditional PI region of macaque monkeys. However, they accept the possibility that part of the region defined as PL may be part of PI_{CL}, as suggested by Gutierrez et al. (1995). Adams et al. (2000), based on their connectivity data, argue that ventral PL should not be included as part of the PI, in spite of the fact that PI_{CL} and ventral PL look neurochemically similar.

Chapter 4
Visual Map Representations in the Primate Pulvinar

It has long been suspected that the pulvinar participates in visual functions because it receives direct projections from the retina (Campos-Ortega et al. 1970) and from the SC (Benevento and Fallon 1975; Lin and Kaas 1979). Additionally, the pulvinar is connected with several visual cortical areas (Campos-Ortega and Hayhow 1972; Ogren and Hendrickson 1976; Benevento and Davis 1977). The retinotopic organization of the pulvinar was studied using electrophysiological techniques in the owl (Allman et al. 1972), capuchin (Gattass et al. 1978a), and macaque monkeys (Bender 1981).

In owl monkeys, Allman et al. (1972) found a single topographic representation of the contralateral visual field in PI, where the central 10° of the visual field was represented dorsally and rostrally and more peripheral parts of the visual field were represented ventrally. The lines of isoeccentricity extended caudomedially and corresponded to the small fiber bundle input that also courses caudomedially through PI. The horizontal meridian divides the nucleus into a rostromedial portion, in which the lower visual quadrant is represented, and a caudolateral portion, in which the upper visual quadrant is represented.

Gattass et al. (1978a) described two retinotopic maps in the pulvinar of the capuchin monkey. The map located in the ventrolateral portion (P_{LV} or P1), which comprises PI and the ventral portion of PL, has a greater representation of the central part of the visual field. The projection of the vertical meridian runs along its lateral border, while that of the horizontal one extends from the dorsal third of the hilus of the LGN to the medial border of PI. The lower quadrant is represented dorsally, while the upper quadrant is represented ventrally. The second map, called $P\mu$ (equivalent to P4), located in the dorsomedial portion of PL, is rotated 90° clockwise around the rostrocaudal axis. The vertical meridian is found at the ventromedial border of this nucleus, and the lower quadrant is represented laterally and the upper quadrant medially. Both projections are restricted to contralateral visual hemifield (Fig. 4.1).

R. Gattass et al., *The Pulvinar Thalamic Nucleus of Non-Human Primates: Architectonic and Functional Subdivisions*, Advances in Anatomy, Embryology and Cell Biology 225, https://doi.org/10.1007/978-3-319-70046-5_4

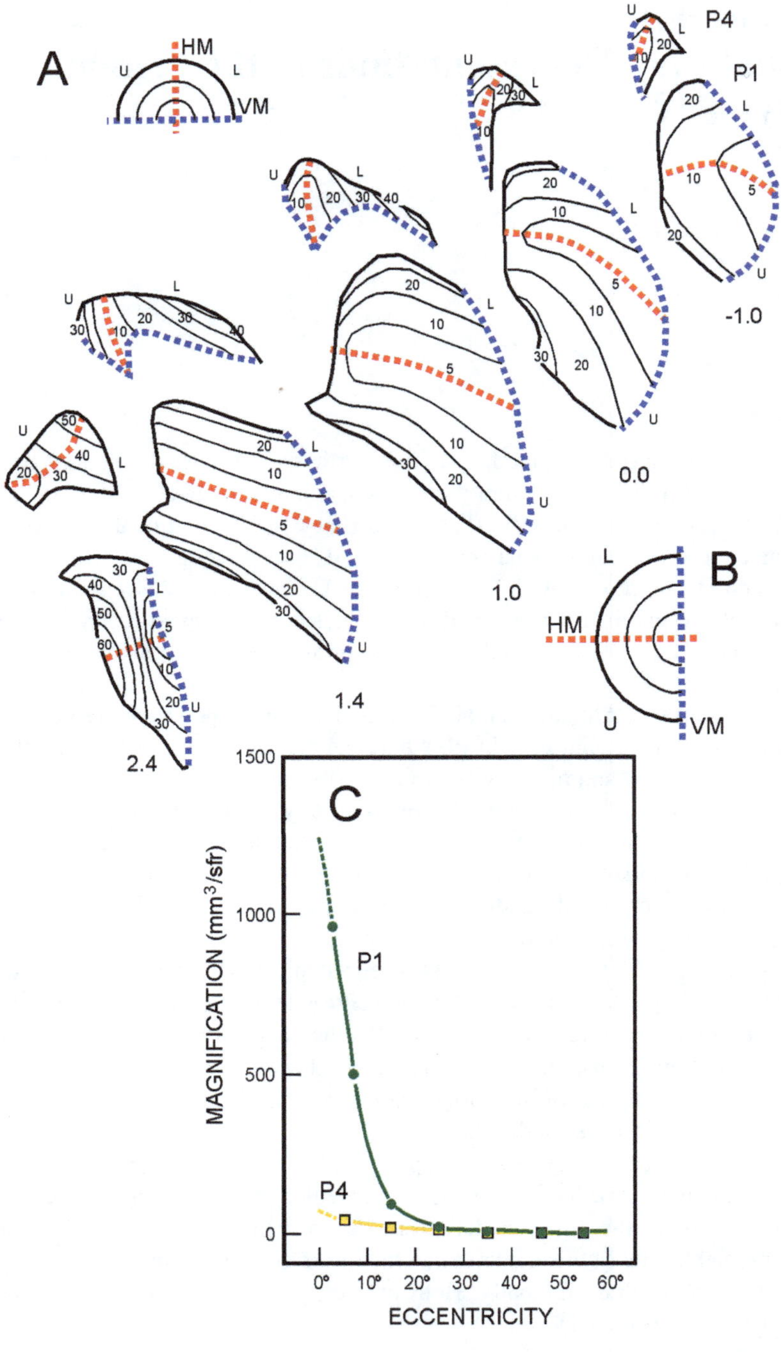

Fig. 4.1 Visual topography of the P4 (Pμ, A) and P1 (P_{LV}, B) sub-regions of the capuchin monkey pulvinar (right hemisphere). The visuotopic organization of the pulvinar is represented in polar coordinates, corresponding to the visual hemifield representations also shown in Panels (**a**) and (**b**).

By recording from clusters of neurons in the pulvinar of the macaque monkey, Bender (1981) also found two representations of the contralateral hemifield. One representation lies mainly within PI but extend into the adjacent PL. The vertical meridian lies in the dorsal and lateral margins of the PI, while the representation of the periphery is found in the medial margin adjacent to the MGN. Central vision is represented laterally and posteriorly. The second representation lies within the PL. The lower quadrant representation lies at its dorsal portion, while the upper quadrant is represented at the ventral half. The horizontal meridian lies in the external margin of the pulvinar. The two maps share a common representation of the vertical meridian.

The visuotopic organization of PI is similar in all three primate species. The vertical meridian and central vision representation are found at the border with LGN, while the peripheral vision representation is found adjacent to the MGN. However, in capuchin and macaque monkeys, this map extends into the adjacent PL. A second visuotopic map has been described in both capuchin and macaque monkeys. However, the lateral map of the macaque monkey is different in extent and visuotopic organization from that of the capuchin Pμ, which is smaller in size and is located more dorsally in the pulvinar.

Petersen et al. (1985), in a behavioral study with macaque monkeys were able to describe, in addition to PI and PL as proposed by Bender (1981), a dorsomedial region (Pdm) of PL, which has a crude retinotopic organization. Other regions exhibiting visual responses, but without clear retinotopic organization, were described in PL by Benevento and Miller (1981). They found neurons in the caudal part of PL, called PLɣ, with large, bilateral receptive fields that were sensitive to changes in luminance levels, were selective to various types of stimulus motion, and exhibited complex binocular interactions. This area seems homologous in location and visual properties to area Pμ of the capuchin monkey described by Gattass et al. (1978a).

Subsequently, Ungerleider et al. (1983, 1984), based on the pulvinar connectivity with V1 and MT, termed the PI and PL maps of the macaque monkey, respectively, the “P1” and “P2” fields. The second map in the capuchin monkey, named Pμ by Gattass et al. (1978a), was located dorsally to PI and was subsequently named “P4” by Adams et al. (2000). P4 may be at least in part coextensive with Pdm (Petersen et al. 1985). Ungerleider et al. (1984) described a third field, “P3,” in the pulvinar characterized by rich connectivity with area MT. It is located

Fig. 4.1 (continued) U and L denote upper and lower visual fields, respectively. The antero-posterior plane is indicated in millimeters at the bottom of the sections. (**c**) The magnification factor (in cubic millimeter per stereoradian, mm^3/str) of P1 and P4 as a function of eccentricity. Note that P1 has a much larger representation of the central visual field as compared to P4 [modified from Gattass et al. (1978a)]

posteromedially in PI but also includes small adjacent portions of PL and PM that lie dorsal to the brachium of the SC (see also Standage and Benevento 1983). P3 does not seem to have a well-defined visuotopic map like its neighbor P1, although it has yet to be mapped electrophysiologically.

Chapter 5
Connectivity of the Pulvinar

Pulvinar connectivity has been studied using a variety of neuroanatomical tracing techniques in both New and Old World monkeys.

PM, for example, has been shown to be interconnected with vast regions of the brain, including the frontal, parietal, and temporal cortices. Several authors have thereby suggested that the PM can be considered a multimodal integrative center (Trojanowski and Jacobson 1974, 1976; Asanuma et al. 1985; Baleydier and Morel 1992). Despite its importance, we will not address PM in depth in the present work. Instead, we will focus on PI and PL due to their important and direct relation to visual function, as evidenced by their visuotopic organization and their connectivity with several visual areas.

As we suggested above, the differentiation of the pulvinar along primate evolution took place upon a relatively rigid chemoarchitectonic scaffold. With increasing cortical size, we argue that the pulvinar developed new functional subdivisions in order to effectively interconnect and interact with the cortex. Therefore, a fully developed chemoarchitectonic scaffold is readily observed in smaller primate species. However, limited neocortical growth and specialization also limited cortical-pulvinar connectivity. This, in turn, hindered pulvinar functional specialization, including the development of visual maps.

Figure 5.1 illustrates the capuchin pulvinar organization and subdivisions based on chemoarchitectural (left) or connectivity/visuotopy (right) criteria. Several cytoarchitectonic subdivisions have been proposed by different authors (Friedmann 1912; Walker 1938; Olszewski 1952). Some of them, as the one originally proposed by Friedmann (1912), have a good correspondence with the chemoarchitectonic subdivisions observed in small and large nonhuman primates. However, their borders do not correlate with the P1–P4 fields described by Ungerleider et al. (1983, 1984, 2014) and by Adams et al. (2000).

In Fig. 5.2 we present a summary of the reciprocal pulvinar-cortical connections based primarily on our own work (Adams et al. 2000; Soares et al. 2001; Ungerleider et al. 2014; Gattass et al. 2014, 2015) but also on Gutierrez et al.

R. Gattass et al., *The Pulvinar Thalamic Nucleus of Non-Human Primates: Architectonic and Functional Subdivisions*, Advances in Anatomy, Embryology and Cell Biology 225, https://doi.org/10.1007/978-3-319-70046-5_5

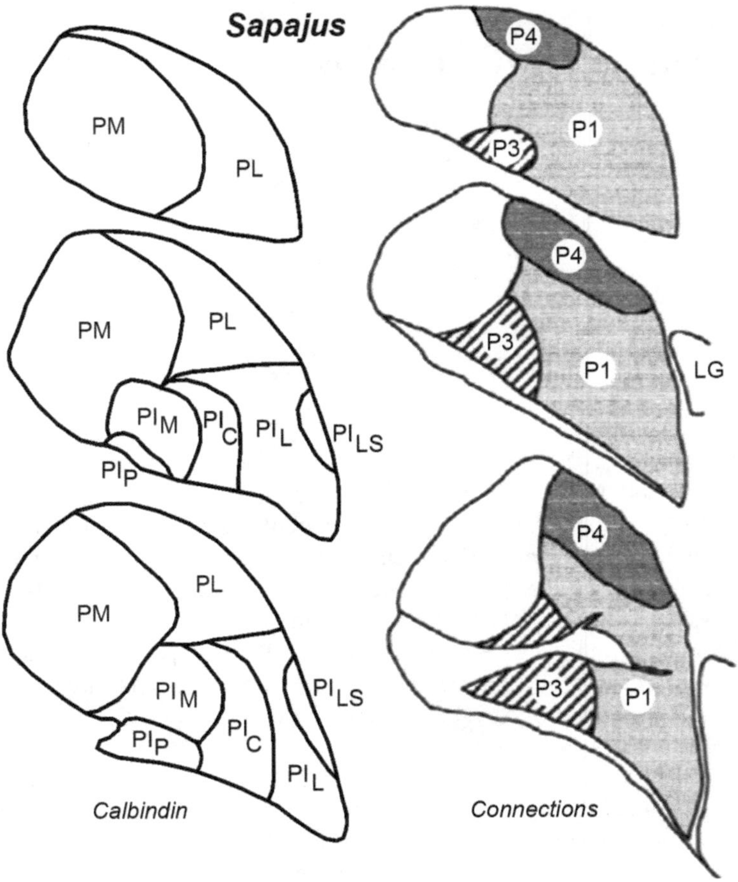

Fig. 5.1 Correspondence between chemoarchitecture and connectivity in the capuchin monkey pulvinar (right hemisphere). Chemoarchitectural subdivisions are based on calbindin immunoreactivity, while the P1–P4 subdivisions are based on the projection fields between the pulvinar and other brain regions, including the cortex. See text for details

(2000), Bridge et al. (2016), and Bourne and Morrone (2017). PI and PL are reciprocally connected with the ventral and dorsal streams of information processing. PI_M receives direct input from the retina and the SC and projects to several areas of the dorsal stream, namely, areas MT, MST, and FST. PI_C and PI_P are also connected with these areas and also with the crescent of MT (MTc). PI_L and PL_V are strongly interconnected with cortical areas of the ventral stream. PL_D projects to the inferior parietal cortex (IPC) and the dorsolateral prefrontal cortex (DLPF). Gutierrez et al. (2000) were able to define PL_D by tracer injections in the IPC and DLPF, but not with injections in the superior temporal gyrus (STG). PM_M and PM_L project to the temporal and parietal cortices, while PM_M also projects to DLPF, orbitofrontal cortex (OFC), and the amygdala. PM_L labeling was found after

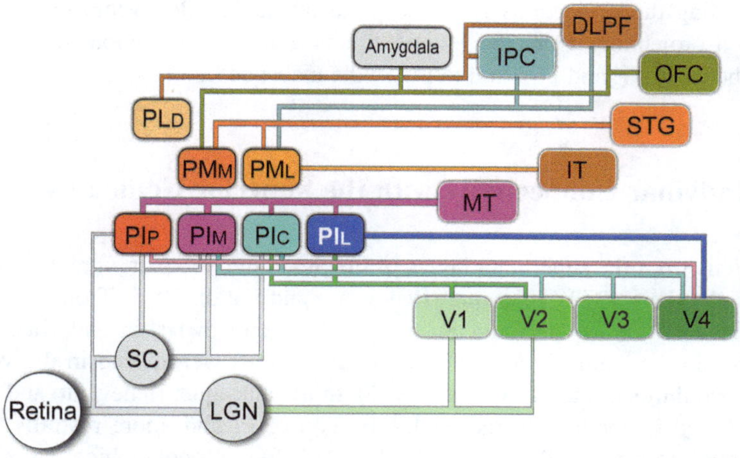

Fig. 5.2 Different cortical areas are richly interconnected with specific subregions of the pulvinar. Rectangles representing cortical regions are depicted on the right, whereas subcortical structures are represented by rectangles and circles on the left of the figure. Colored lines illustrate bidirectional anatomical connections. Projections originating in the retina and SC (grey lines) are unidirectional. PI_P, PI_M, PI_C, and PI_L posterior, medial, caudal, and lateral subdivisions of the inferior pulvinar, respectively. PM_M, PM_L medial and lateral subdivisions of the medial pulvinar, respectively. PL_D dorsal subdivision of the lateral pulvinar. *LGN* lateral geniculate nucleus, *SC* superior colliculus, *STG* superior temporal gyrus, *IT* inferotemporal, *IPC* inferior parietal cortex, *OFC* orbitofrontal cortex, *DLPF* dorsal–lateral prefrontal cortex

injections in the prefrontal and IPC and, to a lesser extent, after injections in the STG. PM_M, on the other hand, was labeled after tracer injections in IPC and STG but only sparsely following tracer injections in the prefrontal cortex (Gutierrez et al. 2000). PM_L and PL_D showed overlapping labeling following paired IPC/prefrontal injections, but not following paired IPC/STG injections. Based on their connectivity data, Gutierrez et al. (2000) conclude that PL_D may be related to visuospatial functions, whereas PM may also be involved in auditory processing.

5.1 Pulvinar Connectivity with the Retina and Pretectal Nuclei

The evidence regarding the existence of direct retinal projections to the pulvinar is conflicting. However, some studies in macaque monkeys and baboons indicate a weak direct retinal projection to the PI subdivision (Campos-Ortega et al. 1970; O'Brien et al. 2001). Cowey et al. (1994) showed that HRP injections in the retinorecipient region of the PI of macaque monkeys were able to retrogradely label Pα and Pβ ganglion cells and an even larger number of Pγ cells in the retina. Recently, Warner et al. (2010) showed the existence of direct retinal synaptic projections to MT relay cells in PIm.

Regarding the projections from the pretectal nuclei, Benevento et al. (1977) described projections from the retinorecipient zone of this region to PL, to the border between PL and PM, and to PO in the macaque monkey.

5.2 Pulvinar Connectivity with the Superior Colliculus

Projections from the superficial layers of the SC to the pulvinar were described in squirrel (Mathers 1972), macaque (Benevento and Fallon 1975; Trojanowski and Jacobson 1975; Benevento and Standage 1983), and owl (Lin and Kaas 1979) monkeys. In the squirrel monkey, the SC projections were found in the ventral-medial two thirds of the PI (Mathers 1971). In the macaque, Benevento and Fallon (1975) found projections to the medial portion of PI and, more rostrally, to the dorsolateral portions of PI. Partlow et al. (1977) found topographically organized projections to most of the PI, with the lower visual field being represented dorsomedially, and the upper visual field being represented ventrolaterally. The peripheral representation was located along the medial border, and the fovea representation was found to be at the dorsolateral angle, adjacent to the LGN.

Lin and Kaas (1979), based on connectivity and architectonic criteria, subdivided the PI complex of owl monkeys into three distinct regions: the central inferior pulvinar (IPc), the medial inferior pulvinar (IPm), and the posterior inferior pulvinar (IPp). Both IPp and IPc receive projections from the SC, but the terminations in IPp are denser than those in IPc (Lin and Kaas 1979). Terminations in IPm from MT are also particularly dense. None of these visual structures project to IPp. Rather, input to IPp appears to originate from the cortex rostral to MT in the temporal lobe.

Benevento and Standage (1983), in addition to the projections to PI, also described projections from the retinorecipient SC layers to three zones in PL. One of the projections lie mainly along the dorsoventral lateral border of subdivisions PLɣ and PLβ, extending ventrally to PLα. A second projection occupies the lateral portion of PLα, and the last one is located medially in PLɣ and PLβ. These PL subdivisions were described by Rezak and Benevento (1979) when studying the organization of pulvinar projections to the primary visual cortex (V1) of the macaque monkey. They found that projections from PL are restricted to the rostral lateral portion immediately adjacent to and cupping PI (PLα). This portion was distinguished from the PLβ, which lies above and adjacent to PLα, and from the caudal portion (PLɣ) which did not project to V1.

In addition to the cortico-pulvino-cortical route, there is a pathway that connects the superficial layers of the SC to the dorsal visual cortices MT and V3, through PI_P and PI_M (Berman and Wurtz 2010; Glendenning et al. 1975; Lyon et al. 2010). Because the superficial SC layers receive direct retinal input, this pathway probably represents a second route from the retina to the visual cortex that bypasses the LGN. The fast transmission time estimated between the SC and MT (Berman and Wurtz 2010) suggests that this brainstem-pulvinar-cortical pathway may be well suited to mediate motion detection, saliency processing, and saccadic suppression and thereby reveal the contribution of the pulvinar to cortical visual processing, perception, and action (Berman and Wurtz 2011).

Inasmuch as brainstem cholinergic inputs suppress zona incerta activity (Trageser et al. 2006), increased vigilance may result in the disinhibition of pulvinar neurons, including the facilitation of transmission along the colliculo-cortical pathway (Trageser and Keller 2004). Trojanowski and Jacobson (1975) also showed direct projection from the LGN to PI and PL in the macaque monkey. Therefore, due to the overall connectivity pattern, the pulvinar may be important to regulate cortico-cortical transmission according to behavioral context.

5.3 Cortical-Pulvinar Connectivity

Large parts of the primate visual cortex exhibit precise visuotopic organization (Gattass et al. 2005, 2015). There are reciprocal and topographically organized visual projections from PI and PL to the striate, prestriate, inferotemporal, and parietal cortices. These projections are observed mainly in cortical layers I, II, and III of area 17 and layers I, III, and IV of areas 18 and 19 (Campos-Ortega and Hayhow 1972; Ogren and Hendrickson 1975, 1976, 1977; Benevento and Rezak 1975, 1976; Trojanowski and Jacobson 1976; Rezak and Benevento 1979). In area 18, the labeling intensity was heavier than that observed for area 17. Pulvinar terminals were found in layer IV, in the lower portion of layer III, and in layer I as alternating zones of densely and sparsely labeled patches (Ogren and Hendrickson 1977; Rezak and Benevento 1979; Wong-Riley 1977). In the dense regions, grains extended vertically from layer IV to layer I in a manner previously reported for somatosensory cortex by Jones et al. (1978). The mean center-to-center distance between the dense zones is greater and more variable than that seen in layer II of area 17 (Ogren and Hendrickson 1977).

The cortico-pulvinar neurons were pyramidal in shape and ranged in size from small to large. In heterotypical cortex, they were found in layers V and VI, whereas in area 17, they were found mainly in layer Vb (Lund and Boothe 1975; Trojanowski and Jacobson 1976; Ogren and Hendrickson 1977).

The topographical organization of the V1-pulvinar projections were described in the marmoset (Spatz and Erdmann 1974), macaque (Campos-Ortega and Hayhow 1972; Ogren and Hendrickson 1976; Ogren 1977), and squirrel (Holländer 1974; Ogren 1977) monkeys. Two regions of the pulvinar receive fibers from V1: PL and PI. Tracer injection within the central visuotopic representation resulted in the labeling of the lateral halves of PI and PL, while peripheral injections showed more medial labeling. The vertical meridian is represented in the architectonical boundary between these two pulvinar regions, consistent with previous electrophysiological recordings (Allman et al. 1972; Bender 1981).

Lin et al. (1974), using horseradish peroxidase (HRP), showed that projections from the pulvinar to MT in the owl monkey were originating mainly from the medial portion of the PI. In the macaque monkey, Standage and Benevento (1983) found overlapping retrogradely filled cells and anterogradely transported terminal grains located exclusively within a crescent-shaped region, which traverses the brachium of the SC to include the PI and PL. The connections between MT and the pulvinar crescent are reciprocal and topographically organized, with the lower visual field represented dorsally and the upper visual field represented ventrally (Soares et al. 2001). There is an expanded representation of central vision located caudally within the crescent, while peripheral vision is represented rostrally (Standage and Benevento 1983).

Using HRP and ^{3}H leucine in combination, Trojanowski and Jacobson (1975) demonstrated that the connections between the superior temporal gyrus (STG) and the pulvinar are topographical and reciprocal. STG projects to and receives projections from the ventromedial PM, while the posterior third is reciprocally connected mainly to ventromedial PL and medial PI. More recently, the connections of the pulvinar with the visual cortex were studied in macaque monkey with tracer injections in areas V1, V2, V4, MT, and PO (Adams et al. 2000; Gattass et al. 2014; Ungerleider et al. 2014). The connectivity of the pulvinar with area V2 was studied by Ungerleider et al. (2014) in macaque monkeys using multiple tracer injections in different eccentricities in V2. Figure 5.3 shows the regions of the pulvinar containing the corresponding labeled summary of seven selected cases. It is possible to observe well-defined topographic maps in P1 and P2, and a cruder map in P4, which may exhibit some degree of segregation between the upper and lower visual field representations. The projections from V2 in these seven cases encompass almost the entire extent of the P1, P2, and P4 fields of the pulvinar. The injection sites in Cases 2, 6, and 10, which were located in the upper visual field representation of V2, led to ventral patches in P1 (Fig. 5.3, A +2.5–A+2.0) and P2 (Fig. 5.3, A +2.0–A+1.0) and to a central patch in P4 (Fig. 5.3, A +2.0–A+1.0). The injections in the lower visual field representation of V2 led to dorsal patches in P1 (Fig. 5.3, A +2.5–A+2.0) and P2 (Fig. 5.3, A +2.0–A+1.0) and to both dorsal and ventral patches in P4 (Fig. 5.3, A +2.0–A+1.0). The patches revealed by the V2 injections show a considerable overlap in all pulvinar fields, suggesting coarser topographic organizations in these fields as a result of convergent input and/or larger receptive fields in the pulvinar, when compared to those in V2 (Ungerleider et al. 2014).

Gattass et al. (2014) studied the connectivity of the pulvinar with V4 in the macaque monkey. Figure 5.4 illustrates the distribution of labeled cells and terminals in the pulvinar and surrounding regions after injections of anterograde and retrograde tracers (HRP, Bis, and 3H) in V4 in one macaque monkey (Case 5 from Gattass et al. 2014). Several clusters of labeled cells and terminals were found in P1 (sections 1–5), P2 (sections 2–6), P3 (sections 3–5), and P4 (sections 3–6).

Figure 5.5 shows projecting cells and terminals in a montage of parasagittal sections after an injection in the upper visual field representation of V4. Note the four patches with labeled cells and terminals, corresponding to the P1, P2, P3, and P4 fields. Note also that the patches corresponding to the projections originating

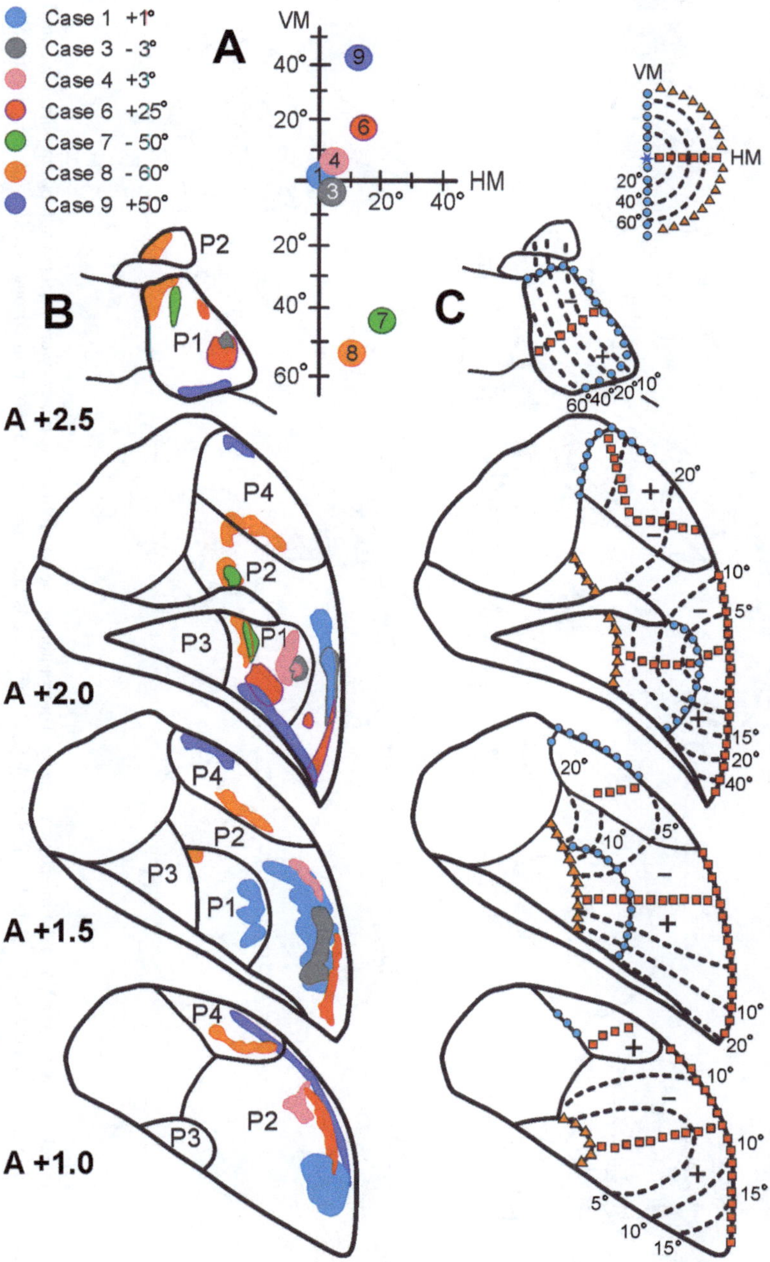

Fig. 5.3 Projections from V2 to the pulvinar. Three topographically organized projection zones (P1, P2, and P4) of the macaque monkey pulvinar revealed by anterograde tracer injections into V2 at different eccentricities for seven selected cases (upper left). The list of the cases and the locations of the receptive field centers in V2 are shown in (**a**). (**b**) Reconstructions of the projection zones shown on coronal sections of the pulvinar from anterior (+2.5) to posterior (+1.0) section planes. (**c**) Representation of the topographical maps in the projection zones P1, P2, and P4 of the pulvinar plotted on the corresponding sections shown in (**b**) [modified from Ungerleider et al. (2014)]

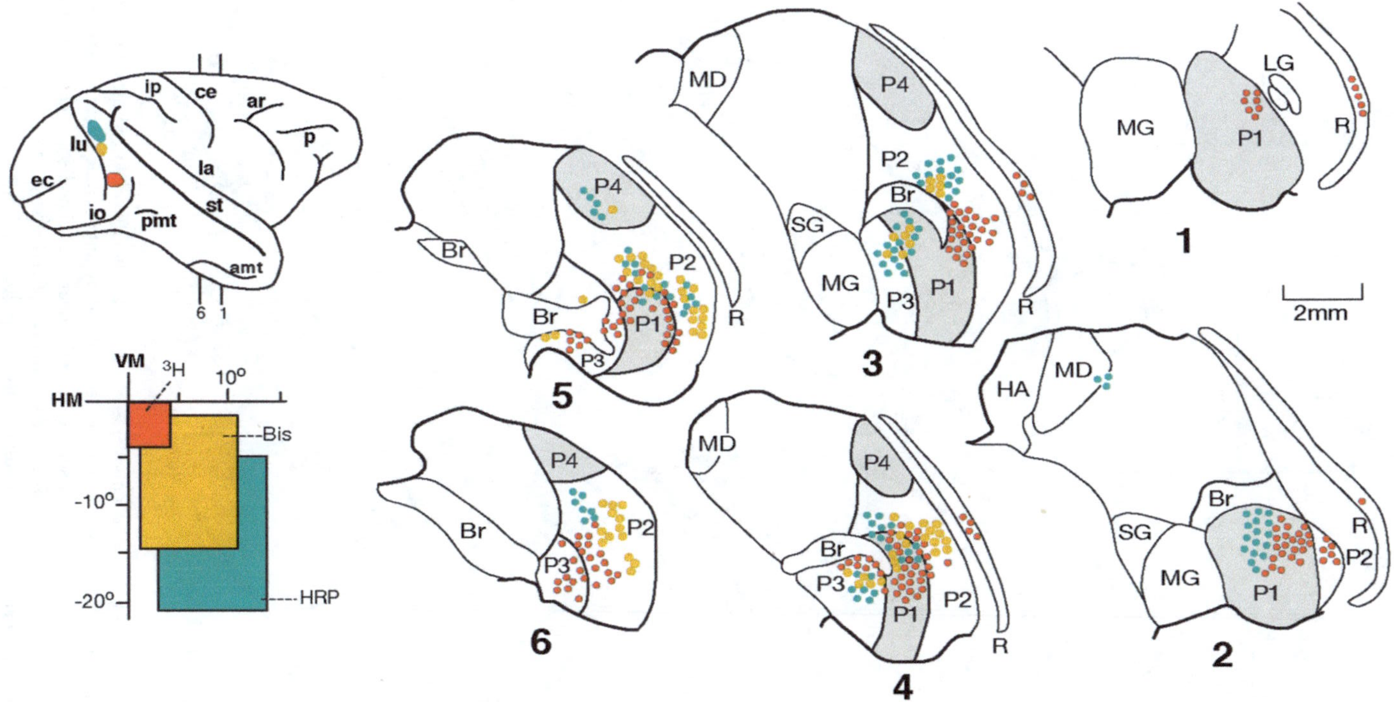

Fig. 5.4 Connections of the pulvinar with area V4 in the macaque monkey. Three anterograde and retrograde tracers were injected at topographically defined locations in V4, as depicted in a lateral view of the right hemisphere (left). Labeled cells and terminals are shown in coronal sections of the pulvinar and surrounding areas (right). See text for details [modified from Gattass et al. (2014)]

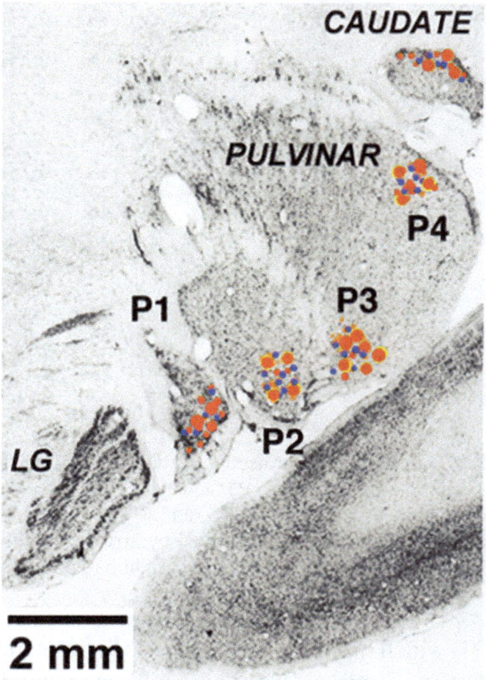

Fig. 5.5 Connections of the pulvinar with area V4 (parasagittal sections of the left hemisphere). Retrograde labeled cells (red-orange concentric icons) and anterograde labeled terminals (blue dots) were found at the topographically corresponding locations of the P1–P4 projection zones of the pulvinar after HRP injection in V4. Note also the projections to and from the caudate to V4 [modified from Gattass et al. (2014)]

from a tracer injection site in the upper visual field representation of V4 are located ventrally in fields P1–P3 and dorsally in P4 (Gattass et al. 2014).

Adams et al. (2000) showed that projections from the pulvinar to V1 and V2 in macaque monkeys overlap in two separate fields that are in register with the visual field maps of P1 and P2. In some, but not all, cases, an additional projection field was found from P3 to V2. However, we did not observe reciprocal projections from V2 to P3 in all cases. MT projecting cells were also found in P1 and P2 but were mainly concentrated in the medial portion of P3. Adams et al. (2000) also showed extensive projections from P2 to V4 but sparser projections from P1 and still sparser from P3. Our current scheme shows that V2 projecting neurons terminate in P2, P3, and P4, similar to the projection field of area V4.

Immunohistochemical studies in macaque, capuchin, and squirrel monkeys have revealed five similar subdivisions of the pulvinar, which include all of PI but which also encompass parts of PL and PM. These regions have been named PI_P, PI_M, PI_C, PI_L, and PI_{LS} (Cusick et al. 1993; Gutierrez et al. 1995; Gray et al. 1999; Adams et al. 2000; Soares et al. 2001). The similarities in the chemoarchitectonic subdivisions contrast with the distinct connectivity and the different visuotopic organizations found in the pulvinar among these species.

In the capuchin monkey, Soares et al. (2001) were unable to clearly segregate P1 from P2 based on their connectivity pattern with areas V1, V2, MT, and V4, in spite of the great chemoarchitectonic similarities between macaque and capuchin monkeys. Areas V2 and V4 in the capuchin monkey have preferential connections with

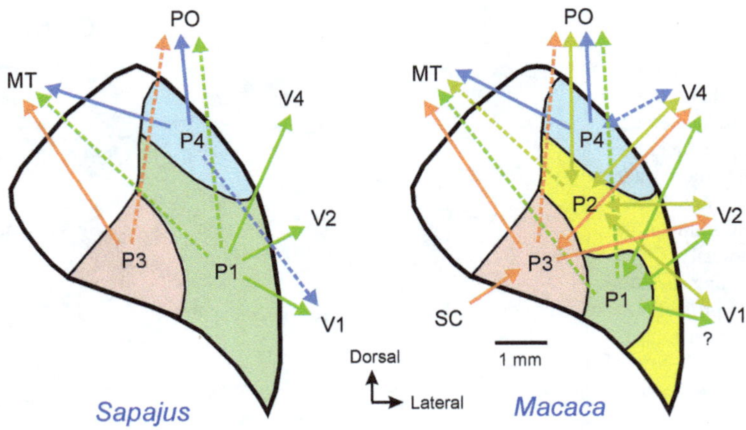

Fig. 5.6 The three projection fields (P1, P3, and P4) of the capuchin pulvinar and the four projection fields (P1, P2, P3, and P4) of the macaque pulvinar revealed after tracer injections in cortical areas V1, V2, V4, PO, and MT. Strong and weak projections between the pulvinar and these five regions are represented by arrows (continuous and dashed, respectively) [Left—modified from Soares et al. (2001) and right based on Colby et al. (1988) and Gattass et al. (2014)]

P1, which may correspond to the ventrolateral complex described in this species by Gattass et al. (1978a) and could correspond to both P1 and P2 described in the macaque monkey. A similar segregation was described by Cusick et al. (1993) and Stepniewska and Kaas (1997), who established that the subdivisions of PI that receive ascending projections from the SC are distinct from the portion of the nucleus that projects to area MT.

Figure 5.6 compares the three projection fields (P1, P3, and P4) of the capuchin monkey pulvinar with the four projection fields (P1, P2, P3, and P4) of the macaque monkey pulvinar, focusing on their projections to and from visual areas V1, V2, V4, PO, and MT. Projections from the SC are also illustrated in this figure.

Kaas and Lyon (2007) have further proposed that the pulvinar nuclei could be segregated into two groups related to the two streams of visual information processing, namely, the ventral and dorsal streams for object and spatial vision, respectively (Mishkin and Ungerleider 1982). According to this proposal, the pulvinar nuclei would provide cortico-pulvinar-cortical interactions that would enable the spread and integration of information both within each visual stream and across streams, in addition to relaying visual information from the SC, via P3, to the dorsal stream cortical areas (Ungerleider and Christensen, 1977).

There are two feedforward projections to V2: one from the lateral/inferior pulvinar and the other from V1. Inasmuch as neither the pulvinar nor V2 can be visually activated following V1 removal, either or both of these inputs could serve as a neuronal driving source to V2 (Marion et al. 2013). Reversibly inactivating the PL in the galago (a prosimian primate) was found to prevent supra-granular V1 neurons from responding to visual stimulation (Purushothaman et al. 2012). Conversely, reversible, focal excitations of the lateral pulvinar were found to increase fourfold the visual responses in coincident V1 receptive fields and shift partially overlapping V1 receptive fields toward the topographic representation of the excitation site (Purushothaman et al. 2012). Excitation of PL after LGN lesions activated supra-granular layer V1 neurons. If these results also hold in other primates, then PL would be in a strategic position to control and gate information outflow from V1 during changes of state or attention (Purushothaman et al. 2012; Sherman and Guillery 2002). Consistent with this potential role of the pulvinar in regulating the effects of spatial attention, deactivation of PL causes spatial attention deficits in monkeys (Desimone et al. 1990). Finally, joint recordings in V4 and PL of macaque monkeys performing a visual attention task show synchronized activity between the two structures (Saalmann and Kastner 2011).

Chapter 6
Reestablishing the Chemoarchitectural Borders Based on Electrophysiological and Connectivity Data

The neurons projecting from the P1 and P2 regions of the pulvinar to area V2 overlapped those neurons projecting to area V1 in all cases studied by Ungerleider et al. (2014). Interestingly, in the two animals with retrograde tracer injections in areas V2, MT, and V4, the neurons in the pulvinar projecting to area V2 overlapped more extensively with those projecting to area V4 than with those projecting to area MT. The region with the most extensive overlap of cells projecting to areas V2 and V4 occurred in PI_{CL} of the P1 and P2 fields. There was minimal overlap between the projection zones from the pulvinar to V2 and MT. This was due to the fact that after retrograde tracer injection in area MT, the labeling found in P1 and P2 was sparse. In the P1 and P2 fields, neurons were found to project to both V1 and V2. However, in these cases, we did not observe any double-labeled pulvinar neuron projecting both to V2 and V4 or to both V2 and MT.

The P3 field was found to send projections to visual areas V2, V4, and MT, but not to V1. Each chemoarchitectonic region within the P3 field was observed to project differentially to each of these cortical areas. In all cases, projections to V2 arose from PI_{CM}, and in three of these cases, projection neurons were also seen in PI_M. Projections to area V4 arose from PI_P and PI_{CM} in all cases, while area MT received projections from all P3 subdivisions. In the two monkeys with retrograde tracer injections in V2, MT, and V4, we observed dense overlapping projection zones from within P3's PI_{CM} subdivision to areas V2 and V4. The overlap between pulvinar fields projecting to MT and V2 was minimal mainly because the neurons projecting to area MT were quite few. Unlike the overlapping pulvinar projection fields to areas V1 and V2, in all of the monkeys studied with retrograde tracer injections in MT and V4, there was a clear interdigitation of labeled neurons. This was because the strongest pulvinar projections to MT and V4 seemed to originate, respectively, from subregions PI_M and PI_P/PI_{CM} (Fig. 6.1). Adams et al. (2000) suggested that the pulvinar integration of cortical afferents and efferents could take advantage of the lamellar organization of the chemoarchitectonic divisions, where superimposed concentric shells are aligned through their visuotopic organization.

R. Gattass et al., *The Pulvinar Thalamic Nucleus of Non-Human Primates: Architectonic and Functional Subdivisions*, Advances in Anatomy, Embryology and Cell Biology 225, https://doi.org/10.1007/978-3-319-70046-5_6

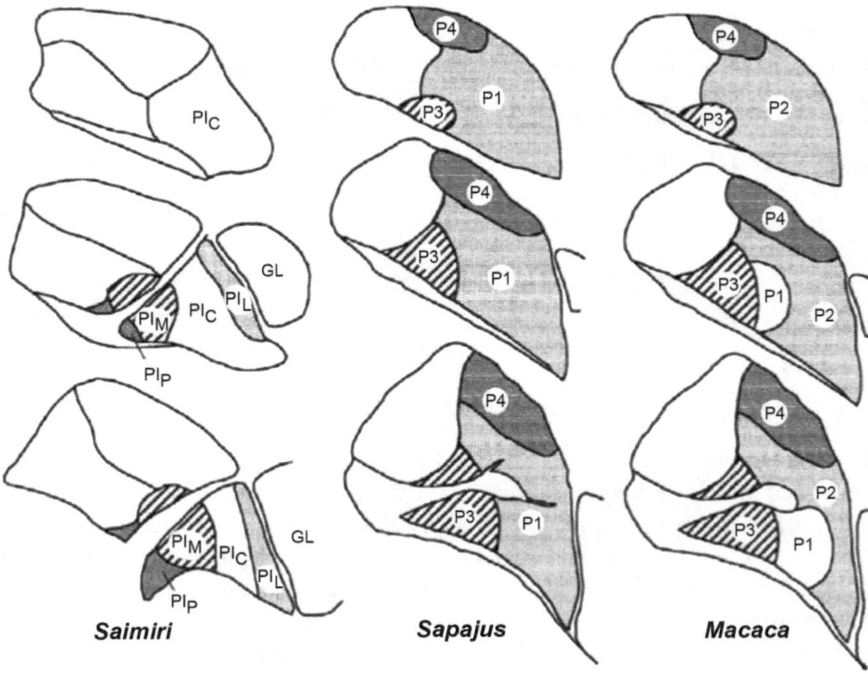

Fig. 6.1 A comparison of the projection fields of the pulvinar (right hemisphere representation) in squirrel, capuchin, and rhesus monkeys. See text for details

This concentric shell structure would allow local topographic integration necessary for spatial visual enhancement or suppression of specific visual information.

Investigating the calbindin immunoreactivity in the macaque monkey pulvinar, Adams et al. (2000) extended the original cytoarchitectonic subdivisions previously described by Olszewski (1952) and offered new insight into pulvinar delimitation and connectivity pattern. PM was easily distinguished from PL by its distribution pattern of both large and small calbindin-containing neurons and by its neuropil disposition and staining. In addition, calbindin staining suggested further partitioning of PI that was not evident in previous studies using only the Nissl stain. The original PI could thereby be further subdivided into PI_P, PI_M, PI_{CM}, and PI_{CL}. Both the PI_P and PI_C zones displayed the strongest calbindin immunoreactivity. The PI_{CL} zone was moderately reactive for calbindin staining, while PI_M was almost devoid of calbindin immunoreactivity. The border between P1 and P2, and the lateral border of P3 with P1 and P2, could be defined using calbindin immunoreactivity. Nissl staining, while being able to identify these borders, revealed an otherwise fairly homogenous staining pattern, which was only interrupted by the brachium of the SC.

The neurons that project to area V1 in New and Old World monkeys were found in the dorsal portion of PI_{CL}, and they overlapped the pulvinar field projecting to area V2 in all animals studied (Stepniewska and Kaas 1997). Additionally, the

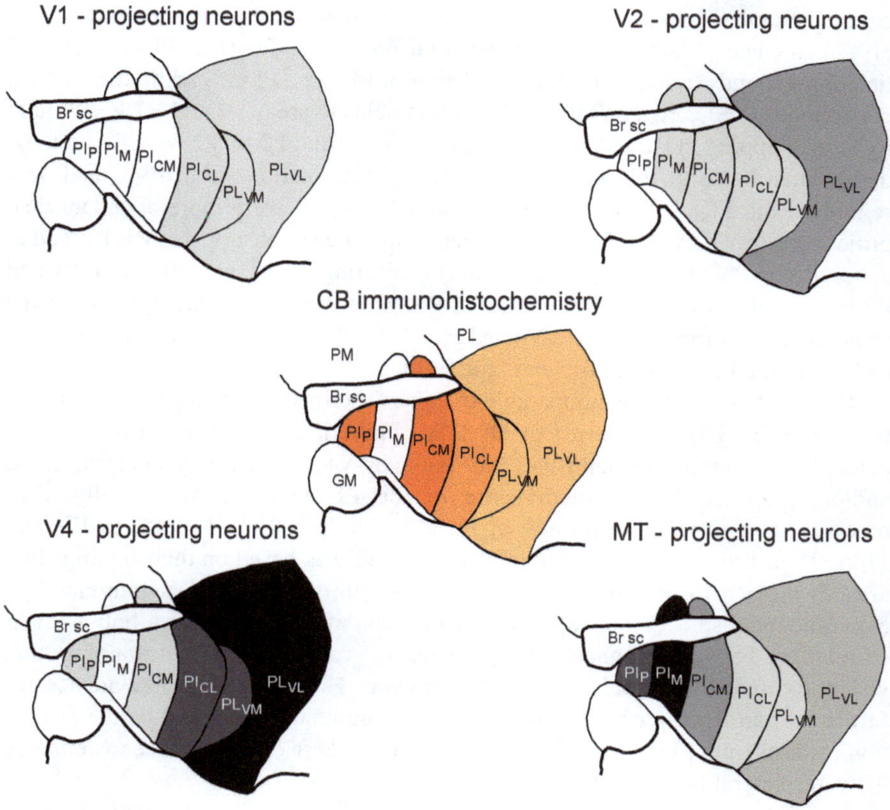

Fig. 6.2 Subdivisions of the pulvinar based on chemoarchitecture (calbindin immunocytochemistry) and its relationship with pulvinar-cortical connectivity. The illustration allows direct correlation of calbindin immunoreactivity with the density of pulvinar-cortical connectivity (gray scale) with visual areas V1, V2, V4, and MT. Areas PL_{VM} and PL_{VL} are labeled PI_L in Figs. 3.2 and 5.2. For details, see text (modified from Adams et al. (2000)]

pulvinar projection zones to areas V1 and V2 fell within the visual field maps of P1 and P2. There was another projection zone to area V2 that was found in P3 of all monkeys studied. The neurons projecting to area MT were found in the ventral part of PM, scattered throughout PL, and in the PI_L, PI_M, PI_{CM}, and PI_{CL} subdivisions. Therefore, in accordance with Ungerleider et al. (1984), the strongest projections to MT originated in P3. Note that the labeled neurons seen in the ventral part of PM could have been the result of an extravasation of the tracer injections into neighboring area FST. The neurons projecting to area V4 were observed in PI_{CL}, PI_{CM}, and PI_P and occupied portions of all three visual field maps (P1, P3, and P4).

Figure 6.2 offers an integrated perspective of pulvinar partitioning based on calbindin immunostaining and the connectivity strength of each of the corresponding pulvinar subdivisions with visual areas V1, V2, V4, and MT. Note the PI partitioning into posterior (PI_P), medial ($PI_{M)}$), central medial (PI_{CM}), and central lateral (PI_{CL}) regions. The P1 field, as described by Ungerleider et al. (1983), includes PI_{CL} and the ventromedial portion of PL (i.e., PL_{VM}). The P2 field corresponds to the ventrolateral portion of PL (i.e., PL_{VL}), while the P3 field corresponds to PI_P, PI_M, and PI_{CM}. Projection fields to areas V1 and V2 were found to be overlapping in P1 and P2, but the projections from P2 to V2 were found to be denser than those to V1. V2 also received light projections from PI_{CM} and, less reliably, from PI_M. Pulvinar projections to V4 and MT were more abundant than projections to V1 and V2. Neurons projecting to V4 were observed in P1, P2, and in subdivisions PI_P and PI_{CM} of P3. Neurons projecting to MT were found in P1 and P2 but mainly in the PI_M subdivision of P3. Note the interdigitated nature of the projection fields from P3 to area V4 and area MT (PI_P and PI_{CM} subfields projecting to V4 and the PI_M subfield projecting to MT).

These subdivisions are comparable to those described by Cusick and colleagues (Cusick et al. 1993; Gutierrez et al. 1995; Gutierrez and Cusick 1997). Using calbindin immunohistochemistry and pulvinar-V1 connectivity pattern, these authors proposed further subdivisions of the PI complex. Additionally, they renamed the ventrolateral portion of PL (i.e., the lateral shell of PI) as PI_L and $PI_{L\text{-}S}$. Their proposed nomenclature change for PI was based on their findings that area V1 projects to all of the subdivisions of the ventral portion of the pulvinar. Our experimental results do not support their new classification, since we find no clear correlation between connectivity patterns or calbindin immunostaining and visuotopic function organization in the pulvinar. However, there is evidence for further PI partitioning based both on calbindin immunostaining and on the projection patterns from PI to areas MT and V4. Notably, these criteria suffice to delineate the ventrolateral border of P3.

Stepniewska and Kaas (1997) used calbindin immunohistochemistry in New World and Old World monkeys to delineate the subdivisions of the pulvinar, particularly to further subdivide the PI (Table 1.1). Their delineation takes into account that a portion of PL is chemoarchitecturally similar to PI and thereby extends the former above the brachium of the SC. Their results fit better with the functional maps described by Gattass et al. (1978a) and Bender (1981) and also with our chemoarchitectonic work (Soares et al. 2001).

Chapter 7
Visual Topography of the Pulvinar Projection Zones

Injection of tritiated amino acids in V2 revealed topographically organized projections targeting the P1, P2, and P4 regions of the pulvinar. The topographic map in P1 was originally described by Gattass et al. (1978a) in the capuchin monkey and subsequently by Bender (1981) in the macaque monkey. According to these authors, P1 is located in the PI but also includes a small portion of the immediately adjacent PL. The peripheral visual field is represented anteriorly in the medial portion of PI, while the central visual field is represented more posteriorly in the medial portion of PL. The vertical meridian is represented on the lateral edge of the nucleus, while the horizontal meridian is represented obliquely from lateral to medial across the nucleus and tilted slightly downward. The upper field is represented ventrally, while the lower field is represented dorsally.

The P1 map resembles a first-order transformation of the visual field (Allman and Kaas 1971). In the macaque monkey, the lateral portion of P1 seems to follow the P2 organization. It contains a representation of the peripheral visual field, which is located in the anterior portion of the nucleus, and a representation of the central visual field, which is located in the posterior portion of PL (Bender 1981). P2 and P1 share the representation of the vertical meridian, while P2's horizontal meridian representation is a continuation of P1's, in a way that the foveal region is represented at the lateral border of the pulvinar. Thus, the P2 map resembles a second-order transformation of the visual field (Allman and Kaas 1971). Gattass et al. (1978a) described a topographically organized region in dorsal pulvinar that seems to be equivalent to the P4 field described by Adams et al. (2000). P4 has a complex topographic arrangement. The representation of the vertical meridian is located on the dorsal edge of P4, while the representation of the horizontal meridian exits the dorsal edge and divides P4 into dorsal and ventral portions. The representation of the upper visual field occupies the dorsal and anterior P4, while the representation of the lower visual field is located ventral and posterior, adjacent to the lower field representation of P2. The P4 map resembles a distorted first-order transformation of the visual field (Allman and Kaas 1971).

R. Gattass et al., *The Pulvinar Thalamic Nucleus of Non-Human Primates: Architectonic and Functional Subdivisions*, Advances in Anatomy, Embryology and Cell Biology 225, https://doi.org/10.1007/978-3-319-70046-5_7

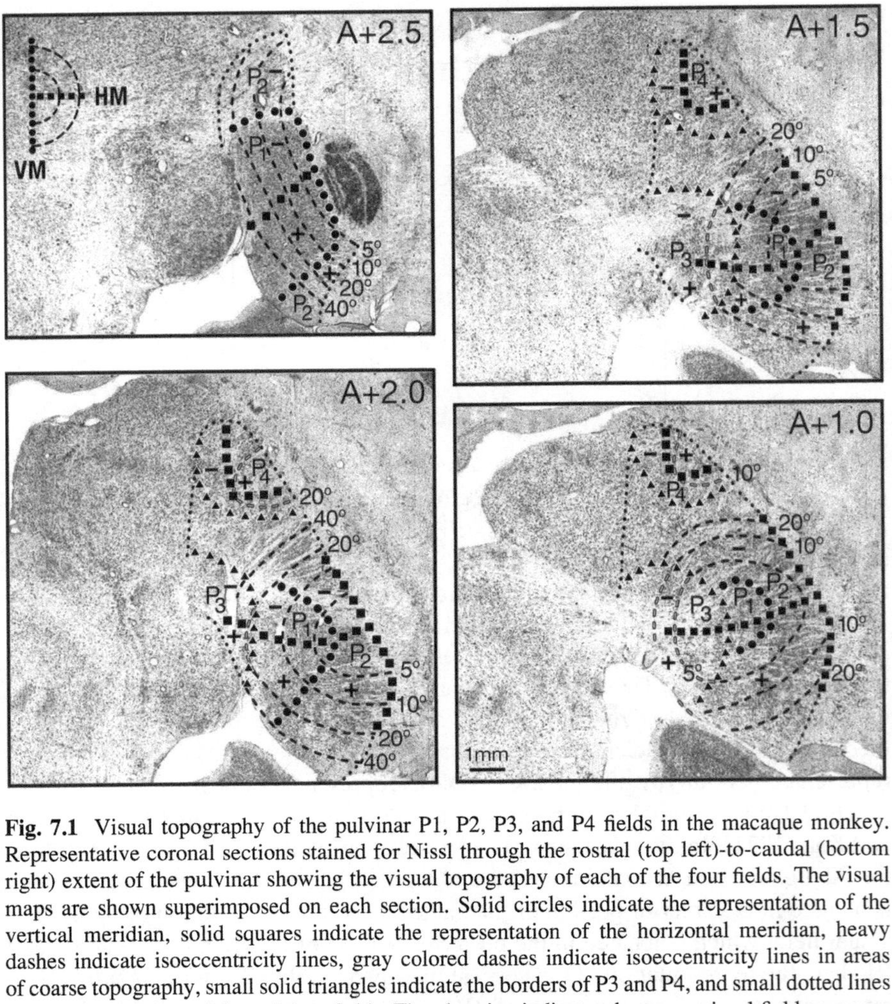

Fig. 7.1 Visual topography of the pulvinar P1, P2, P3, and P4 fields in the macaque monkey. Representative coronal sections stained for Nissl through the rostral (top left)-to-caudal (bottom right) extent of the pulvinar showing the visual topography of each of the four fields. The visual maps are shown superimposed on each section. Solid circles indicate the representation of the vertical meridian, solid squares indicate the representation of the horizontal meridian, heavy dashes indicate isoeccentricity lines, gray colored dashes indicate isoeccentricity lines in areas of coarse topography, small solid triangles indicate the borders of P3 and P4, and small dotted lines indicate the borders of the pulvinar fields. The plus sign indicates the upper visual field representation and the minus sign indicates the lower visual field representation. The sections are spaced 0.5 mm apart, and they do not reach the caudal extent of the pulvinar. Abbreviations as in Fig. 2.1. Scale bars = 1 mm. (Modified from Gattass et al. 2014)

Figure 7.1 shows the visuotopic maps of the macaque monkey pulvinar (P1–P4) charted onto Nissl-stained sections. The P1 and P2 maps are based on the work by Bender (1981) and Ungerleider et al. (1983, 1984). The estimate of P3's borders was guided by the work of Ungerleider et al. (1984) and Gattass et al. (2014). The dorsal border of P3 (i.e., the portion above the brachium of the SC) was adjusted to be compatible with the distribution of calbindin immunoreactive neurons presented in previous work (Adams et al. 2000; Gattass et al. 2014). The estimate of P4's border was also guided by the works of Adams et al. (2000) and Gattass et al. (2014).

Chapter 8
Comparative Pulvinar Organization Across Different Primate Species

Immunohistochemical studies have revealed five similar pulvinar subdivisions (PI_P, PI_M, PI_C, PI_L, and PI_{LS}) in the macaque, capuchin, and squirrel monkeys (Table 1.1 and Fig. 3.3), which include the entire PI but which also encompass parts of the PL and PM (Cusick et al. 1993; Gutierrez et al. 1995; Gray et al. 1999; Adams et al. 2000; Soares et al. 2001). The similarities in chemoarchitecture contrast with the distinct connectivity patterns and the different visuotopic organizations found across species. In the capuchin monkey, Soares et al. (2001) were unable to clearly segregate P1 from P2 based on pulvinar connectivity with V1, V2, MT, and V4, as it is the case in the macaque monkey (Ungerleider et al. 1983). This contrasts with the fact that capuchin and macaque monkeys share a very similar chemoarchitecture. Areas V2 and V4 in the capuchin monkey have preferential connections with the P1 field, which may correspond to the ventrolateral complex of the pulvinar described by Gattass et al. (1978a) and would correspond to both P1 and P2 of the macaque monkey, as described by Ungerleider et al. (1983). A similar partitioning was described by Cusick et al. (1993) and Stepniewska and Kaas (1997), who also established that the subdivisions of PI that receive ascending projections from the SC are distinct from the PI subdivision that projects to area MT. Inasmuch as the PI (P1, P2, and P3) is the only tecto-recipient region of the pulvinar (Partlow et al. 1977), the function of its connections with V4 may include modulating tectal input to this cortical area.

R. Gattass et al., *The Pulvinar Thalamic Nucleus of Non-Human Primates: Architectonic and Functional Subdivisions*, Advances in Anatomy, Embryology and Cell Biology 225, https://doi.org/10.1007/978-3-319-70046-5_8

Chapter 9
Response Properties of Pulvinar Neurons Studied with Single-Unit Electrophysiological Recordings

Mathers and Rapisardi (1973) studied the visual and somatosensory responses of the squirrel monkey pulvinar, where they described visual neurons in subregions PL, PI, and PM. Most neurons exhibited a definable receptive field, while only a few responded to diffuse illumination. Approximately twice as many neurons in PI were responsive to light compared to neurons in PL or PM. Nearly all neurons with identifiable receptive fields responded to visual stimulation within 25° of the fovea, on the hemifield contralateral to the recording electrode. The majority of the visual units were responsive to some form of moving stimulus, and some exhibited direction or orientation selectivity. Most visual neurons were monocularly driven and exhibited receptive fields of at least 100 square degrees in area. Mathers and Rapisardi (1973) also found somatosensory neurons in PL. Most of these units exhibited continuous peripheral receptive fields, though a few of these neurons could be bilaterally activated.

In order to systematize our electrophysiological findings and to enable a coherent presentation of the data, we have classified the neurons recorded in the pulvinar according to their functional properties (Gattass et al. 1978a, b). The units were thereby assigned to eight different categories or groups, as summarized in Fig. 9.1.

The first tier of this classification segregates the pulvinar neurons into either static or dynamic units. Neurons classified as static showed a brisk response to stationary stimuli presented over their receptive fields and a similar or weaker response to moving stimulus. In contrast, dynamic units showed poor or no response to stationary stimuli but a brisk response to moving stimuli. Dynamic units predominated (75%) over static ones (25%). About 15% of the units could not be categorized as either static or dynamic and were thereby designated as "unclassified."

Generally, the dynamic units were tuned for stimulus velocity. Static units responded tonically (58%) or, less frequently, phasically (42%) to stimulus onset or offset. In contrast, dynamic units always responded phasically to such stimulus transients.

R. Gattass et al., *The Pulvinar Thalamic Nucleus of Non-Human Primates: Architectonic and Functional Subdivisions*, Advances in Anatomy, Embryology and Cell Biology 225, https://doi.org/10.1007/978-3-319-70046-5_9

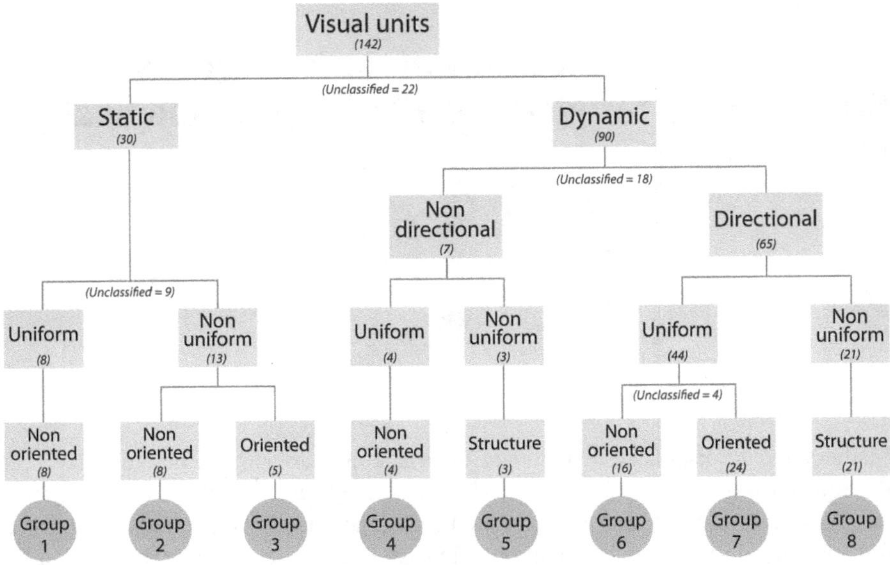

Fig. 9.1 Classification of pulvinar neurons based on their functional response properties measured with electrophysiological recordings. When the data available did not permit a reliable classification of a unit, it was included under a separate heading (i.e. 'unclassified'). The number of units classified under each heading is shown in parenthesis [modified from Gattass et al. (1979)]

9.1 Neurons Classified as "Static"

Static units can be further subdivided on the basis of their receptive field organization. Units classified as "uniform" showed similar response properties throughout the extent of their receptive fields. In contrast, units classified as "nonuniform" exhibited more complex receptive field structure, as outlined below.

9.1.1 Uniform Non-oriented (Group 1)

The "uniform non-oriented" units were nonselective to stimulus orientation and displayed somewhat homogeneous response properties across their receptive field. However, their receptive field borders were not always definable, and some units responded to diffuse illumination. Figure 9.2 illustrates an example of such a unit, which responds with a tonic *on* discharge during stimulus presentation. Note that the response to a 7° spot varies in magnitude depending on the stimulation site within the receptive field.

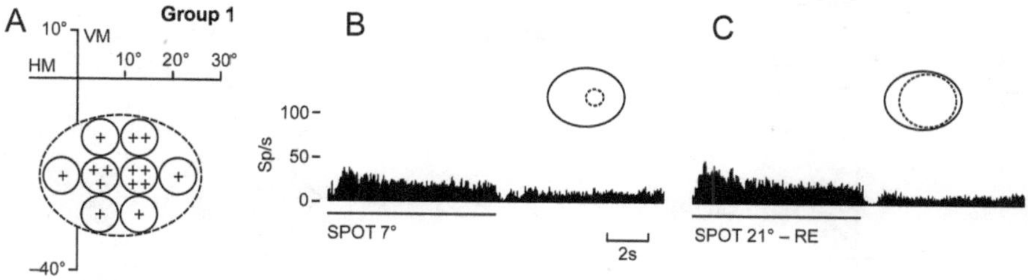

Fig. 9.2 Group 1 unit isolated in P1. This unit gives an *on*-tonic response to stimuli presented anywhere within its RF. In (**a**), the number of crosses indicates the relative magnitude of the response to a 7° spot. Note that the response magnitude does not vary appreciably when either a 7° spot (**b**), a 21° spot (**c**) or diffuse light (not shown) is presented to the RF. The continuous line below the post-stimulus time histograms (*psth*) indicates stimulus duration. (**b**) and (**c**) represent two *on-off psths* of the cumulative number of events that occurred in each of 256 bins following the onset (*on*) and interruption (*off*) of the stimulus presentation. The time span covered by each bin is adjustable. Each histogram represents the cumulative acquisition of 30 trials. Abbreviations used in this and in the other figures: *VM* vertical meridian, *HM* horizontal meridian, *Sp*/*s* discharge rate in spikes per second [modified from Gattass et al. (1979)]

9.1.2 Nonuniform Non-oriented (Group 2)

Neurons classified as "nonuniform non-oriented" were similar to Group 1, except that their receptive field subregions exhibited distinct functional properties. Namely, these neurons responded with either excitation or inhibition depending on the portion of the receptive field being stimulated. For the subset of pulvinar neurons studied in Gattass et al. (1979), responses to stimulus onset or offset could always be evoked by stimulating the receptive field center. Interestingly, visual stimulation on the receptive field periphery was always phasic, regardless if the response to receptive field center stimulation was tonic or phasic.

9.1.3 Nonuniform Oriented (Group 3)

The Group 3 "nonuniform oriented" neurons distinguish themselves from Group 2 by their selectivity to visually oriented stimuli presented within their receptive field. The responses to static stimuli presented to the receptive field center were predominately excitatory and tonic, even though inhibitory or phasic responses could also be observed, especially when stimulating the receptive field surround. Indeed, we observed nonuniformities in the functional organization of these receptive fields. The responses reflected different degrees of center vs. surround interaction, where center-surround antagonism was usually predominant. An example of a Group 3 unit is illustrated in Fig. 9.3. Note the tonic sustained response when stimulating the receptive field center (Fig. 9.3-A_2) and the phasic response when

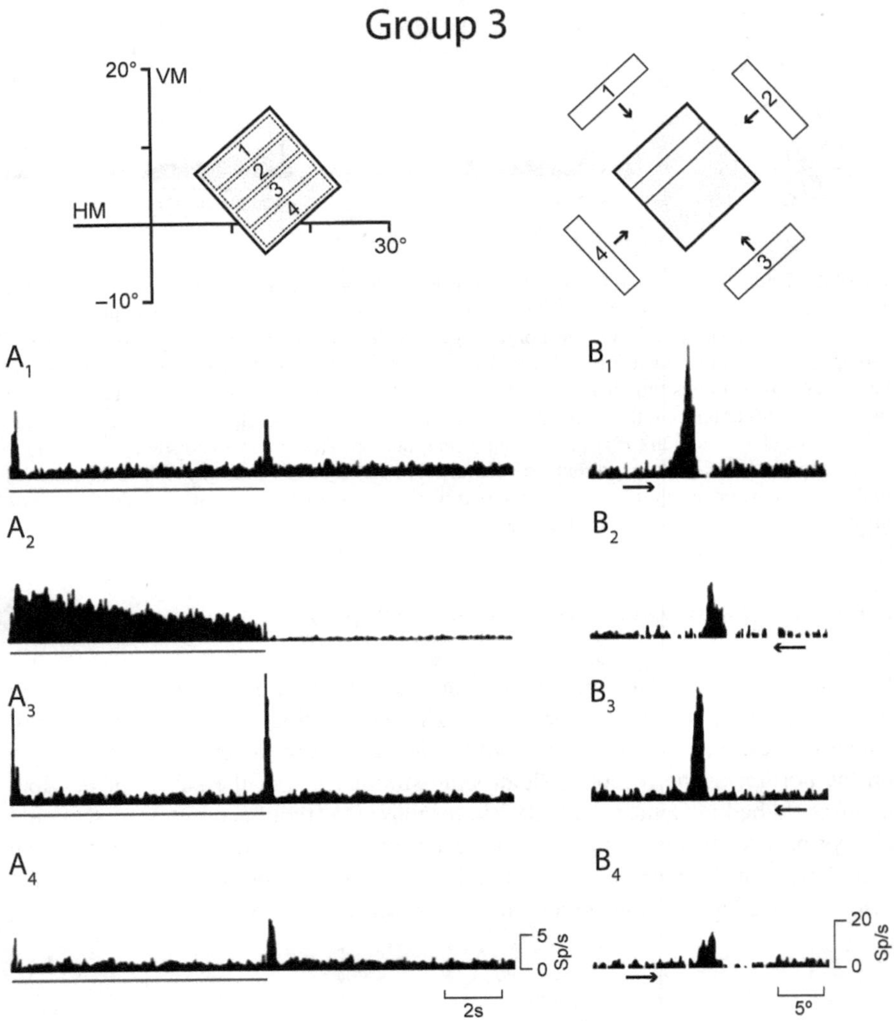

Fig. 9.3 Response characteristics of a P1 single unit (Group 3). (A_1), (A_2), (A_3) and (A_4) illustrate *on-off psths* corresponding to neuronal responses to a stopped slit (2.5° wide) presented in four distinct regions of the RF (top left panel). Note the tonic *on*-response in (A_2) contrasting with the phasic *on-off* response to stimulation in other regions of the receptive field. Panels B_1, B_2, B_3 and B_4 illustrate *psths* of the cumulative number of spike events for each direction of stimulus displacement across the screen (top right panel). The plots in (B) correlate neuronal firing with stimulus displacement in the directions indicated by the arrows. Note that the neuron exhibits a preferred stimulus orientation (see Panels B_1 and B_3), but does not show direction selectivity. Stimulus velocity = 13°/s. Data was gathered for 30 trials (Panels A_1–A_4) and for 15 trials (Panels B_1–B_4) [modified from Gattass et al. (1979)]

stimulating the receptive field flanks (Fig. 9.3-A_1, A_3 and A_4). Panels $B_{1\text{-}4}$ illustrate the orientation selectivity of this particular unit.

9.2 Neurons Classified as "Dynamic"

The main criterion for pulvinar neurons to be classified as "dynamic" constituted their poor response to static stimuli being presented over their receptive fields. Dynamic units could be additionally subdivided based on their sensitivity to stimulus motion direction. Within the "dynamic" group, the majority of the units were direction selective. Units that were nondirection selective usually responded to a luminous spot moving along any axis within their receptive field. Interestingly, units classified as "dynamic" were predominately binocularly driven.

9.2.1 Nondirectional Uniform Non-oriented (Group 4)

Only a small proportion of the "dynamic" units were found to be nonselective for either stimulus direction or orientation. Neurons exhibiting these response characteristics were evenly distributed as presenting a uniform (Group 4) or a nonuniform (i.e., structured; see Group 5) receptive field organization. The neurons showing uniform responses discharged briskly when a spot of light crossed the borders of their receptive fields, including when the stimulus swept at high velocities. However, sustained tonic responses could be elicited using "jerky stimulus movements."

9.2.2 Nondirectional Nonuniform with Structure (Group 5)

The pulvinar neurons with nonuniform receptive fields (Group 5) contrast with those having uniform receptive fields (Group 4) in two ways: Group 5 neurons have smaller receptive fields and show a preference for low stimulus velocities, compared to those neurons classified as Group 4. The work of Gattass et al. (1979) reported on only three units belonging to Group 5. Two of them had a receptive field with a center-surround organization, and one had subregions within its receptive field that were selective to different stimulus properties. Figure 9.4 illustrates a Group 5 neuron with center-surround receptive field organization. The presentation of a static stimulus restricted to the receptive field center produced a weak phasic response (Fig. 9.4b). An annular stimulus sparing the receptive center elicited no response, while it also blocked the neuron from discharging during the simultaneous presentation of a center stimulus. Note that the unit exhibits no orientation or direction selectivity (Fig 9.4c). The small difference in response amplitude observed for the vertical (1–6) as compared to the horizontal (3–4) stimuli may

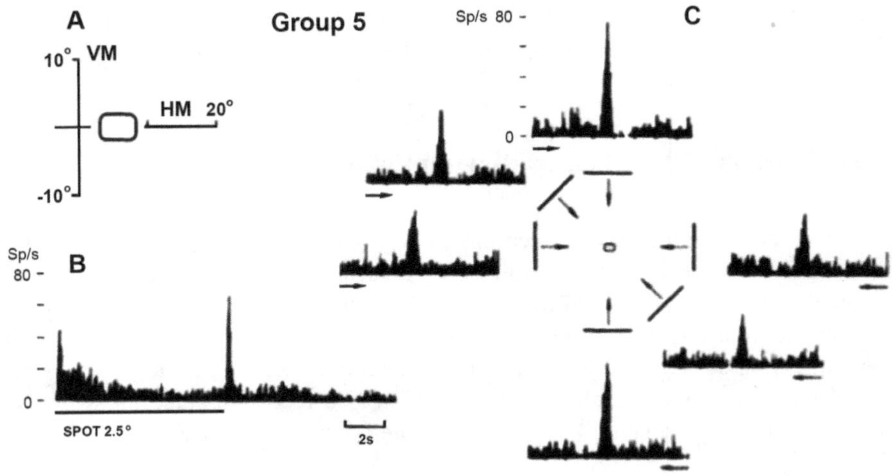

Fig. 9.4 Example of non-directional unit showing surround suppression (Group 5). (**a**) Example unit isolated in P4 and possessing a rectangular receptive field located at the level of the horizontal meridian, near the fovea. (**b**) Phasic ON-OFF response evoked by a 2.5° diameter spot presented at the center of the receptive field. (**c**) PSTHs of the responses of the same unit obtained when a full slit (0.75° wide) is displaced across its receptive field in the directions indicated by the arrows (stimulus velocity = 9°/s, 20 trials) [modified from Gattass et al. (1979)]

be attributed to the larger area of the RF being activated when the slit was oriented parallel to the long horizontal axis of the receptive field.

9.2.3 Directional Uniform Non-oriented (Group 6)

Neurons exhibiting direction selectivity were the most abundant visual units found in the pulvinar. These neurons usually preferred slow stimulus velocities. They could be subdivided into two broad categories: units exhibiting uniform receptive fields either with (Group 7) or without (Group 6) orientation selectivity, as well as units exhibiting nonuniform receptive fields (Group 8). Group 6 neurons were characterized by the fact that they responded equally well to either a spot of light or to a broad set of oriented stimuli displaced across their receptive fields (Fig. 9.5).

9.2.4 Directional Uniform Oriented (Group 7)

Group 7 neurons distinguished themselves from Group 6 units by the presence of orientation selectivity. The vast majority of these cells actually exhibited bidirectional responses. They thereby elicited only weak responses to a spot of light displaced across their receptive field. Accordingly, they were much more narrowly

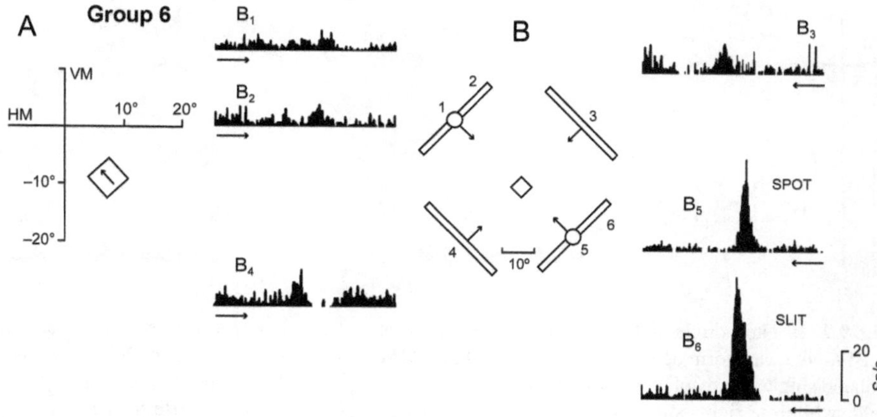

Fig. 9.5 Directional uniform non-oriented unit (Group 6). (**a**) Receptive field of the unit isolated in P1. (**b**) The PSTHs illustrate the unidirectional response of the unit to a 1° wide full slit (stimulus velocity = 7°/s, 20 trials). Similar results were obtained when the slit was substituted by a spot [modified from Gattass et al. (1979)]

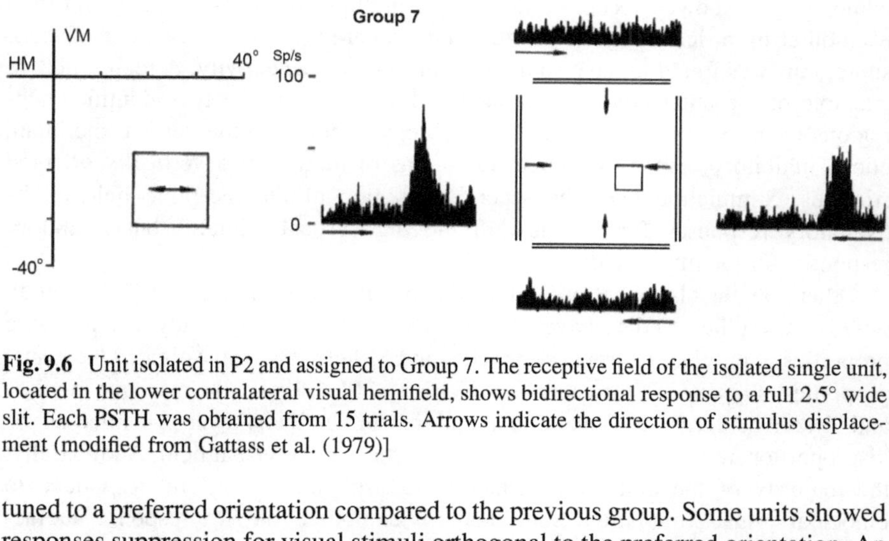

Fig. 9.6 Unit isolated in P2 and assigned to Group 7. The receptive field of the isolated single unit, located in the lower contralateral visual hemifield, shows bidirectional response to a full 2.5° wide slit. Each PSTH was obtained from 15 trials. Arrows indicate the direction of stimulus displacement (modified from Gattass et al. (1979)]

tuned to a preferred orientation compared to the previous group. Some units showed responses suppression for visual stimuli orthogonal to the preferred orientation. An example of a Group 7 unit is illustrated in Fig. 9.6.

9.2.5 *Directional Nonuniform with Structure (Group 8)*

Units classified as "direction nonuniform with structure" (Group 8) had basically two types of receptive field organization. The simplest type of organization exhibited receptive fields with a single responsive region surrounded by inhibitory

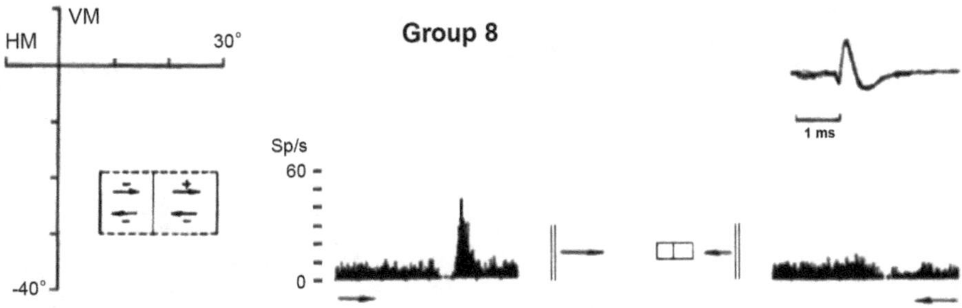

Fig. 9.7 Isolated single unit exhibiting a complex receptive field organization (Group 8) recorded in P1 (spike waveform depicted on the top right). PSTH containing the responses of 20 trials of the isolated single unit displaying a bidirectional inhibitory response in the left portion of the receptive field, while the right portion of the receptive field gave an excitatory response to the stimulus moving in one direction and a inhibitory response to the stimulus moving in the opposite direction. The stimulus was a full 2° wide slit moving at a velocity of 9°/s (modified from Gattass et al. (1979)]

flanks and was usually selective to a single direction of stimulus motion. The more complex type showed receptive fields with multiple subregions interacting with each other in intricate ways to produce a neuronal response. At least one of these subregions was found to have some form of direction selectivity. A representative example of the latter type is illustrated in Fig. 9.7. Note the two identified subregions of its receptive field. The first subregion, closer to the vertical meridian, elicits inhibitory responses to both directions of motion of a vertically oriented stimulus. Stimulation over the second subregion of the receptive field elicits excitatory responses for a rightward moving vertical stimulus but inhibitory responses for the opposite direction.

Other than the classification presented above based on the work of Gattass et al. (1979), few other studies have attempted to systematically study the response properties of pulvinar neurons. Benevento and Miller (1981) investigated the visual properties of neurons in the caudal subdivision of PL (PLɣ) in the macaque monkey and described large, unflanked, bilateral receptive fields, which seemed to be disproportionately representing the central portion of the visual field. Additionally, the majority of the units were sensitive to stimulus motion and responded to binocular visual stimulation. Some neurons exhibited complex response interactions within different subregions of their receptive fields, while others responded to stimuli moving toward or away from the center of gaze.

A comparison of the different types of units found in the pulvinar with those described in the various hierarchical stages of visual processing leads us to an interesting question: what is the functional significance of units in the pulvinar showing properties similar to those described at different levels of the visual processing pathway? If we consider the pulvinar as a link between the geniculostriate and retinotectal systems, the presence of receptive fields showing various degrees of complexity is in accordance with an associative or integrative function and therefore enables this thalamic structure to participate in circuits involved in perceptual selection. This role would also help to explain the preservation of form discrimination in both cats and monkeys after removal of striate and peristriate cortices (Nakamura and Mishkin 1986). The presence of neurons in the pulvinar with complex receptive fields and the dependence of their visual responses on the animal's arousal state (see below) could help to explain the severe deficit produced by pulvinar lesions on discrimination tasks that require a high degree of visual attention (Ward et al. 2002). Patients with pulvinar lesions show deficits in spatial information coding for the contralateral visual hemifield. Specifically, these patients have difficulty localizing stimuli in the affected visual space. These difficulties extend to the binding of visual features that are dependent on spatial information (Ward et al. 2002). Thalamic neglect in humans is rare, and severe attentional deficits that occur due to pulvinar lesions typically do not persist for longer periods. However, a milder cognitive deficit, which consists in slower orienting responses to the contralesional hemifield, is found in some patients and may be a residual form of thalamic neglect (Danziger et al. 2001–2002; Rafal and Posner 1987).

Chapter 10
Modulation of Pulvinar Neuronal Activity by Arousal

In contrast to electrophysiological recordings in the early visual cortex, neuronal activity in the pulvinar is particularly sensitive to anesthesia. Accordingly, under deeper levels of anesthesia, Gattass et al. (1978b) were unable to obtain consistent physiological responses in the pulvinar of the capuchin monkey. Anesthesia level was accessed, for example, by EEG recording in the parieto-occipital region, which typically displayed slow-wave oscillatory patterns associated with drowsiness or initial stages of sleep. In the absence of intentional sensory stimulation, pulvinar neurons could be characterized by spontaneous low-frequency rhythmic bursts of spiking activity. However, multisensory stimulation capable of arousing the animal from deeper anesthesia levels could reestablish the necessary neuronal dynamics and switch the pulvinar into an active state. Under these conditions, cortical slow-wave activity was substituted by a higher-frequency oscillatory pattern associated with arousal. Two types of transitions in pulvinar activity pattern could be observed when arousing the animal with multisensory stimulation (e.g., somatosensory or auditory stimulation). The first type consisted in a profound shift in neuronal firing pattern where, after either somatosensory or auditory stimulation, pulvinar single units changed their dynamics from low-frequency rhythmic activity to higher-frequency rhythmic activity (Fig. 10.1). Note that the interspike interval (ISI) distribution that arises during epochs of multisensory stimulation is compatible with the induction of gamma oscillations (~40 Hz) for the single unit being recorded. Neurons undergoing this type of transition appeared to respond exclusively to visual stimuli, despite the fact that stimulation with other sensory modalities influenced their activity. For example, there was a clear temporal relationship between stimulation onset and neuronal activity for visual stimulation, but not for stimuli of other sensory modalities.

The second type of state transition taking place in the pulvinar affected solely the firing rate level of the neurons instead of their firing pattern and generally required more than one modality of sensory stimulation (Figs. 10.2 and 10.3). Typically, these units exhibited fatigue and habituation for sensory modalities other than the

R. Gattass et al., *The Pulvinar Thalamic Nucleus of Non-Human Primates: Architectonic and Functional Subdivisions*, Advances in Anatomy, Embryology and Cell Biology 225, https://doi.org/10.1007/978-3-319-70046-5_10

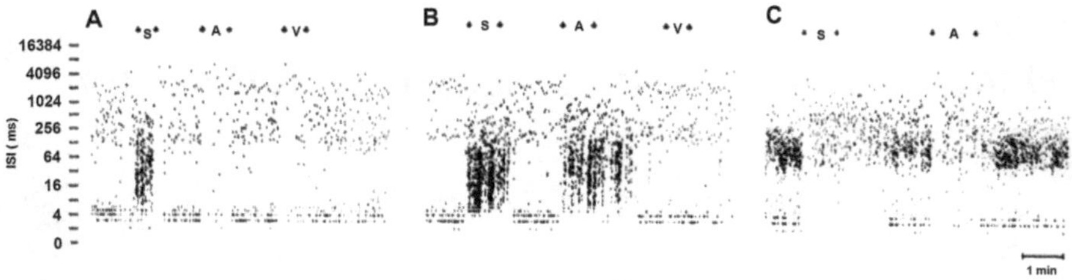

Fig. 10.1 Multisensory stimulation induces change in firing rhythmicity of pulvinar neurons. (**a**), (**b**) and (**c**) show three single units recorded in P1. The single unit depicted in (**a**) increases its firing rate and changes its oscillatory pattern during somatosensory stimulation, but decreases its firing rate during auditory and visual stimulation. The single unit depicted in (**b**) increases its firing rate and changes its oscillatory pattern during somatosensory and auditory stimulation, but decreases its firing rate during visual stimulation. The single unit depicted in (**c**) desynchronizes to somatosensory and auditory stimulation. The firing rate is decreased during stimulation. (Modified from Gattass et al. 1979)

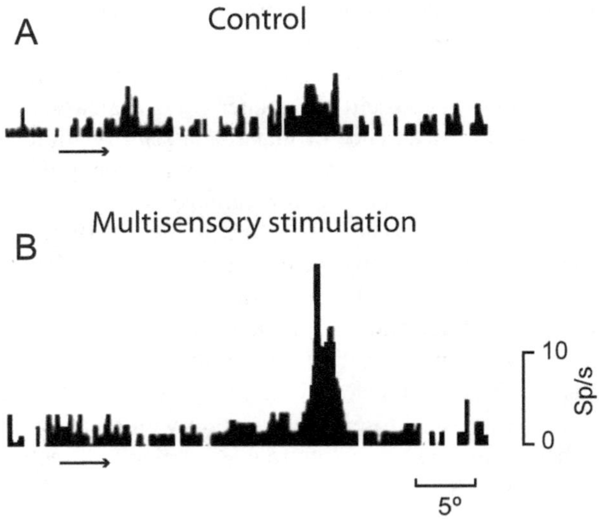

Fig. 10.2 Multisensory stimulation enhances visual responses recorded from a P1 single unit (Group 2 neuron). Visual responses to a moving full slit stimulus (2° wide) are represented in PSTHs before (**a**) and during (**b**) somatic-auditory stimulation. Prior to multisensory stimulation (**a**), the single unit exhibited a rhythmic-cyclic pattern of neuronal activity. Subsequently, when the animal was aroused by multisensory stimulation, the single unit becomes more responsive to the visual stimulation and starts to exhibit direction selectivity. (Modified from Gattass et al. 1979)

visual. Contrary to the pulvinar neurons undergoing the first type of transition, most of these units appeared to be de facto multisensory in a way that they were sensitive to visual, somatic, auditory, and olfactory features of the stimulus. Effective somatosensory stimulation could consist in stimuli applied over a large area of

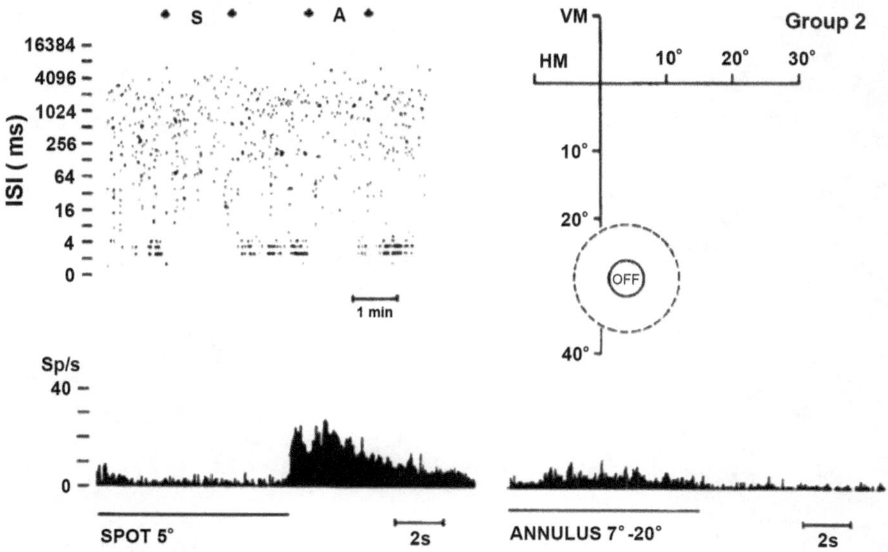

Fig. 10.3 Effect of multisensory stimulation on the visual responses of a single unit recorded in P1 (Group 2). Interspike interval values (ISI, plotted in logarithmic scale) as a function of time (X-axis, linear scale) reveal the rhythmicity of the neuron before, during and after somatosensory (S) and auditory (A) stimulations (Upper left panel, stimulation intervals delimitated by arrows). Each dot represents one spike event. The rhythmicity of the single unit changes from a burst activity, separated by long intervals, to a more homogenous firing. Burst activity shows interspike intervals ranging from 1 to 4 ms, separated by intervals ranging from 10 to 6000 ms. Somatosensory and auditory stimulation prompts the cell to lose its low-frequency rhythmic burst pattern. Note that this unit responds to the static presentation of a bright 5° circle with an OFF sustained response (lower left panel). Stimulation with a bright annulus reveals a poor ON-response (lower right panel). (Modified from Gattass et al. 1979)

the contralateral body surface. Occasionally, ipsilateral stimulation of the fore and hind limb extremities was effective. Flickering or diffuse light was effective as arousing visual stimuli. Finally, complex natural auditory stimuli appeared more effective than pure tones or clicks, despite the fact that we did not seek to parameterize the optimal arousing stimuli.

If it is indeed the case that the pulvinar is highly dependent on arousal levels in order to function adequately, it is reasonable to speculate that the pulvinar itself might be important to "awaken" the cortex (Zhou et al. 2016), especially due to the rich cortico-pulvinar anatomical interconnection. There is evidence that certain types of arousal phenomena, such allocating expectation in time, are capable of activating large portions of the visual cortex (Lima et al. 2011). The pulvinar may thereby function to promote coordinated arousal of large cortical networks.

Chapter 11
GABA Inactivation of the Pulvinar

A direct way to access the pulvinar-cortical interaction is to pharmacologically inactivate the pulvinar and measure the impact on cortical activity. To this aim, we have focused our efforts on recording in cortical visual area V2. The main afferents to area V2 originate in the primary visual cortex (Federer et al. 2009). However, pulvinar subregions PI and PL also provide robust projections to V2 and constitute the major subcortical input to this area. Pulvinar terminal zones align with regions of increased cytochrome oxidase staining in V2, avoiding the pale stripes (Levitt et al. 1995). All V2 stripes receive input from V1. However, the strongest V1 inputs target the V2 pale stripes. This suggests that inputs from V1 and from the pulvinar target distinct V2 modules of visual processing (Sincich and Horton 2002).

We used γ-aminobutyric acid (GABA) injections to inactivate parts of the PI and PL subregions of the capuchin monkey pulvinar (Soares et al. 2001). Our inactivation device included three cannulas placed evenly around a recording electrode. We could thereby monitor pulvinar inactivation using electrophysiological recordings. We usually obtained a 60% reduction in pulvinar neuronal activity immediately after the injection. Recovery of pulvinar activity to levels observed prior to GABA injection could take up to 70 min. Therefore, the corresponding projections from the pulvinar to the cortex were presumably shut down during a significant period. During this period, we were able to record the electrophysiological activity of single neurons in area V2, whose receptive fields matched the topographic representation of the pulvinar inactivation site. Pulvinar inactivation resulted in a myriad of physiological effects in area V2, but two main effects can be here highlighted. The first effect consisted in a general baseline shift in neuronal firing for both spontaneous and stimulus driven activity, which provides further evidence that the

R. Gattass et al., *The Pulvinar Thalamic Nucleus of Non-Human Primates: Architectonic and Functional Subdivisions*, Advances in Anatomy, Embryology and Cell Biology 225, https://doi.org/10.1007/978-3-319-70046-5_11

pulvinar may be involved in the large-scale modulation of cortical arousal. The second effect of pulvinar GABA inactivation involved changes in the receptive field properties of V2 receptive fields, namely, their selectivity for stimulus orientation and direction of motion.

The majority (~67%) of the recorded V2 cells showed changes in baseline activity after pulvinar GABA inactivation, with slightly more V2 neurons undergoing response increment as compared to response reduction (~40% vs. 27%, respectively). GABA can act over several neuronal structures that constitute the complex synaptic glomeruli of the pulvinar. In some cases, it may reduce the postsynaptic depolarization of neurons projecting to the cortex. As a result of pulvinar inactivation, V2 would lose a direct excitatory modulation, which could explain the reduction in its neuronal activity. Alternatively, pulvinar projections could target V2 inhibitory interneurons, causing thereby an increase in general V2 activity after GABA inactivation. The fact that the majority of V2 neurons show a response increase with pulvinar inactivation suggests that inhibitory interneurons may constitute an important component of pulvinar-V2 interaction. Figure 11.1 illustrates a V2 single unit, which showed a reduction in stimulus driven activity to a moving bar after pulvinar inactivation with GABA. Note the substantial reduction in cell response, which starts to recover 83 min. after the injection. The cell activity was recorded before (0) and at different time intervals (21, 83, and 133 min) after injection of GABA. Figure 11.2 illustrates another V2 single unit that increased its baseline activity with GABA inactivation of the pulvinar. Neuronal activity was recorded before (0) and at discrete time intervals (28, 60, and 149 min) after injection. We observed an increment in neuronal activity for both spontaneous and stimulus driven responses. While the stimulus-driven activity begins to normalize at 149 min postinjection, spontaneous activity remains high for reasons which are not completely clear.

An even larger proportion of V2 neurons (~91%) showed changes in orientation or direction selectivity after pulvinar inactivation, where a decrease in selectivity was the most commonly observed effect. This indicates that the pulvinar may play an active role in the neuronal circuit responsible for shaping V2 stimulus selectivity. Note, for example, the successive changes in orientation/direction selectivity that take place during pulvinar inactivation in Fig. 11.2.

Other neurons we studied completely lost their selectivity to orientation or direction but still showed this late rebound effect in the spontaneous activity as seen in Fig. 11.2 and, to a smaller degree, in Fig. 11.1. The physiological mechanisms behind the late rebound effect could be partially explained by the existence of indirect circuits involving other cortical visual areas that receive pulvinar projections and that also project to V2, such as V4 (Zeki and Shipp 1989), MT (Rockland and Pandya 1979), and the inferotemporal cortex (Felleman and Van Essen 1991).

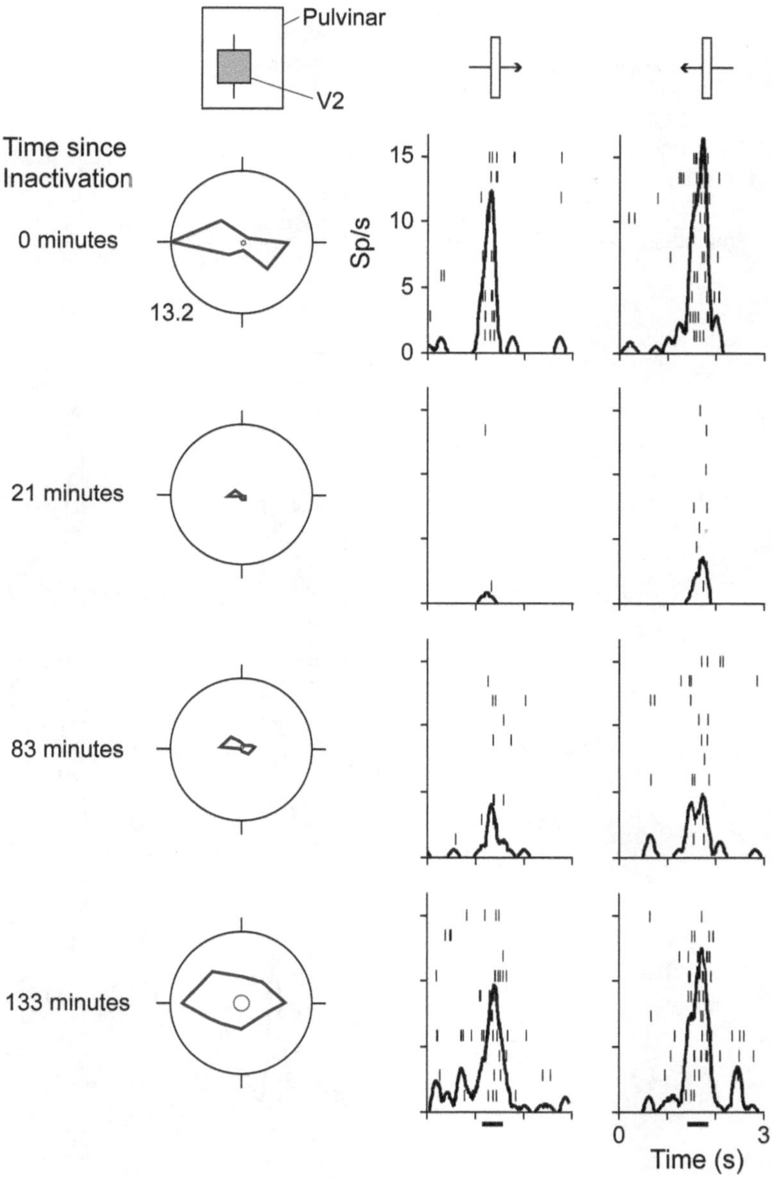

Fig. 11.1 Overall activity decrease of a single V2 unit after GABA inactivation of the pulvinar. Neuronal activity in area V2 was recorded before (0) and at discrete time intervals (21, 83, and 133 min) after GABA injection in the pulvinar. Note the substantial decrease in neuronal activity with GABA, which starts to recover 83 min after the injection. Right: PSTHs of the responses to the preferred stimuli. Left: Polar diagrams displaying the mean response rates for the full set of stimulus directions (45° steps). Dotted-line circles at the center of polar diagrams correspond to the mean spontaneous activity of the cell. The radii of the external circles indicate the maximum recorded response (13.2 spikes/s). Black bars below the PSTHs indicate the time interval during which the stimulus was moving inside the V2 RF. (Modified from Soares et al. 2004)

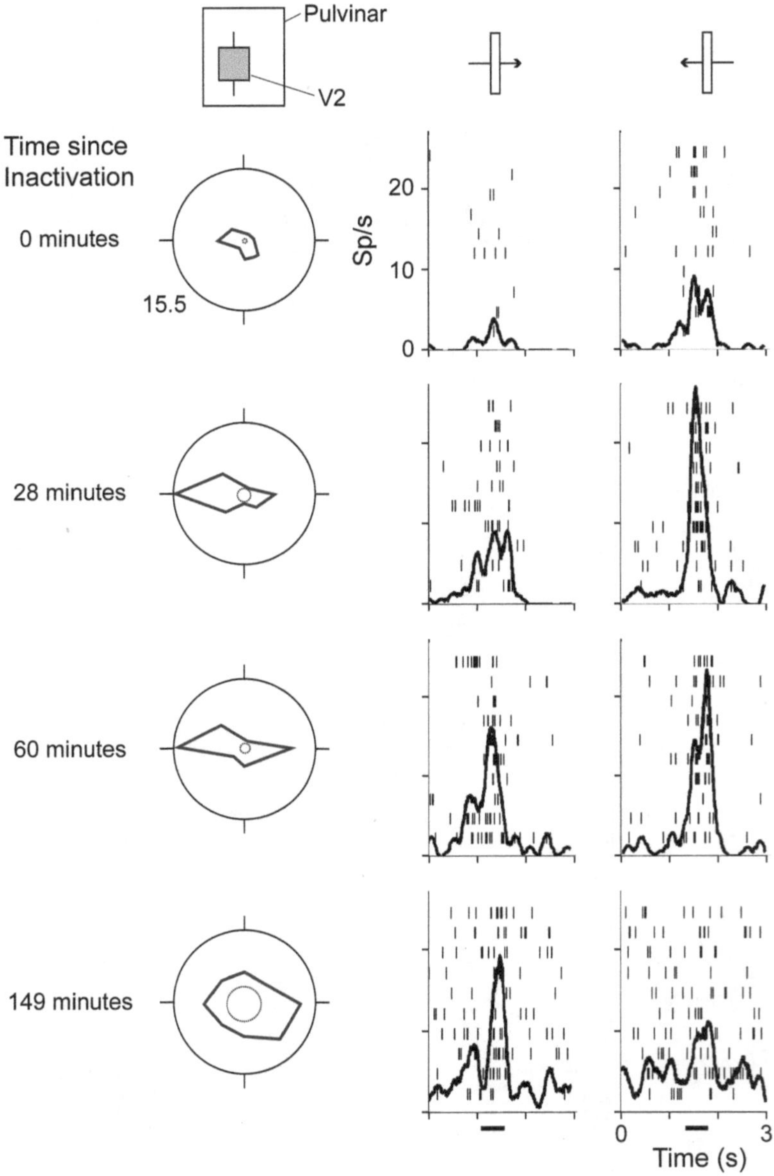

Fig. 11.2 Overall activity increase of a single V2 unit after GABA inactivation of the pulvinar. Neuronal activity in area V2 was recorded before (0) and at discrete time intervals (0, 28, 60, and 149 min) after GABA injection in the pulvinar. Right: PSTHs of the preferred stimuli. Left: polar diagrams corresponding to the tuning curve for direction of stimulus motion (conventions as in Fig. 11.1). Note that the neuron acquired an enhanced selectivity to stimulus direction after GABA injection. Directional index or DI (0 min) = 0.03; DI (14 min) = 0.83. (Modified from Soares et al. 2004)

Chapter 12
The Role of the Pulvinar in Spatial Visual Attention

There are at least two general mechanisms in which the pulvinar seems to be instrumental for spatial visual attention (Fig. 12.1). The first aspect concerns the pattern of pulvinar connectivity with brain regions known to be playing a role in attentional mechanisms. The reciprocal and profuse nature of these projections indicates that the pulvinar is in a strategic position for coordinating activity across large neuronal networks. Therefore, the pulvinar could serve as a hub for brain communication and thereby gate the flow of information across different brain regions. Notably, these connectivity patterns follow a precise visuotopic organization, which is particularly useful for the spatial aspect of visual attention. In addition to its coordinating role, the pulvinar seems to also participate in the selection process of attention. Rafal and Posner (1987) proposed a model for the posterior attention system, which deals with spatial visual attention (Posner and Petersen 1990). They used a very simple attention task that was tested in patients with cortical (i.e., parietal) and subcortical (i.e., pulvinar and SC) lesions. According to these authors, attention is a continuous process with four operational steps: disengage, move, engage, and inhibit (Fig. 12.1). The first step is thereby to disengage from the undesired location and then to redirect attention to the new target location. The previously attended location is now disfavored by the attention system, and the subject responds more slowly at that location than to any other location in the visual field. This tendency to make slower responses to the previously attended location is called "inhibition of return."

Patients with lesions in the parietal cortex show a primary deficit in the disengage operation of spatial visual attention. These patients were unable to move their attention toward the hemifield contralateral to the lesion site (designated as the "bad" visual hemifield). Midbrain lesions involving the SC impaired the ability to redirect the attentional focus. Patients with such lesions lacked an effective "inhibition of return" as they had difficulty moving their eyes and shifting their attention covertly. In patients with pulvinar lesions, one observed a deficit in the ability to hold attention on the target stimulus when competing information was also present

R. Gattass et al., *The Pulvinar Thalamic Nucleus of Non-Human Primates: Architectonic and Functional Subdivisions*, Advances in Anatomy, Embryology and Cell Biology 225, https://doi.org/10.1007/978-3-319-70046-5_12

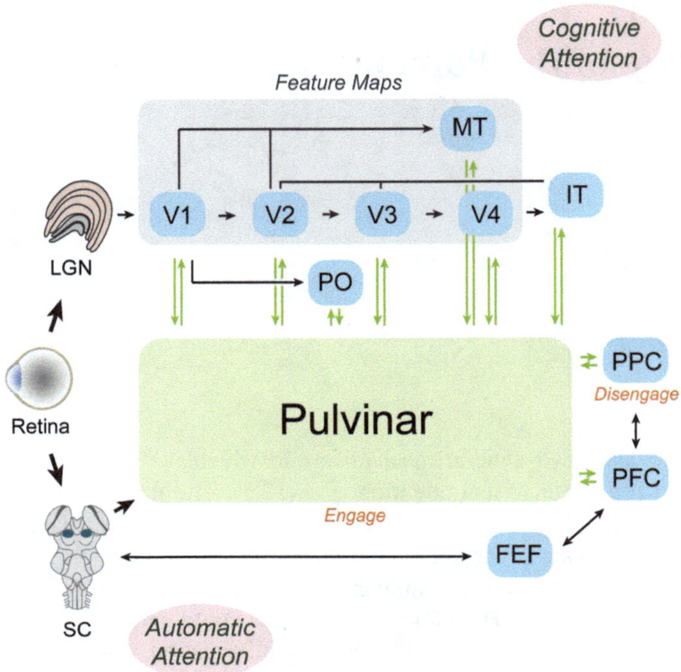

Fig. 12.1 Schematic diagram of the neuroanatomical circuit underlying automatic and cognitive spatial visual attention. An attempt was made to correlate the anatomical structures to the basic attentional mechanisms of disengaging, moving, engaging and inhibiting the focus of attention. V1, V2, V3, V4, PO and MT are cortical visual areas; *LGN* lateral geniculate nucleus, *FEF* frontal eye field, *PFC* pre-frontal cortex, *PP* posterior parietal, *SC* superior colliculus [modified from Gattass and Desimone (1996)]

in the visual field. Therefore, monkeys and patients with pulvinar lesions seem to have a great deal of difficulty filtering out or ignoring irrelevant stimuli that occur at locations other than the one to which they are attending (Desimone et al. 1990; Petersen et al. 1985, 1987). In this sense, these individuals are more distractible and have difficulty engaging their attentional focus (Gattass and Desimone 1996). This scheme incorporates the distinction of at least two attentional mechanisms: one automatic, mediated by the SC, and another central or cognitive, not related to the SC but rather to cortical visual areas with a potential role played by the pulvinar.

From the perceptual point of view, when the eyes are fixed, attention can be covertly directed to another location other than fixation, and information processing at that location will be accordingly enhanced. From the oculomotor perspective, attention at that location can help to determine where to move the eyes (Gattass and Desimone 2014). In normal behavior, the eyes will usually move to that location. Following the eye movement, novel locations, which have not been attended to in the last few seconds, are favored over previously attended locations. Posner's model of attention assumes that attention precedes the eye movement. In this respect, this model is different from the one proposed by Goldberg and Wurst (1972) in which signal programming of the eye movement, coming from the deep layers of the SC, propagates upward to the superficial layers and engages spatial visual attention. The projection of the SC to the cortex, via the pulvinar nucleus, contributes to the visual and attentional properties of cortical visual areas (Gross 1991; Treue and Maunsell 1996).

Although the specific proposals differ, a role for the pulvinar in visual attention has been a recurrent hypothesis (e.g., Chalupa et al. 1976; Gattass et al. 1979; Crick 1984; LaBerge and Buchsbaum 1990; Robinson and Petersen 1992; Olshausen et al. 1993; Shipp 2000). More recent work finds that attentional modulation of activity is pervasive within the pulvinar and as significant as in any area of the cortex (Bender and Youakim 2001). The model presented by Shipp (2003) corroborates this idea. Shipp (2003) reviewed the published data on the topography and connectivity of the pulvinar with the cortex and proposed a simplified, global model for the organization of cortico-pulvinar network. According to this model, connections between the cortex and pulvinar are topographically organized, and as a result, the pulvinar contains four topographically ordered "maps." Shipp (2003) proposed a replication principle of central-peripheral-central projections that operates at and below the level of domain structure. He hypothesized that cortico-pulvinar circuitry replicates the pattern of cortical circuitry, but not its function, thereby playing a more regulatory role instead. The projecting cells in V4 and their termination in the pulvinar suggest that the cortical-pulvinar-cortical connections define a pathway by which deep layers of cortical visual areas affect, via pulvinar, the superficial layers of coupled cortical areas.

In our study of the cortical connections of V4, we found a central-peripheral asymmetry in the projections to the temporal and the parietal cortices (Ungerleider et al. 2008). We concluded that peripheral field projections from V4 to parietal areas could provide a direct route for rapid activation of circuits serving spatial vision and attention, while the predominance of central field projections from V4 to inferior temporal areas could provide the necessary information needed for detailed form analysis for object vision. In the study of the subcortical connections of V4, Gattass et al. (2014) found no evidence for central-peripheral asymmetry. Instead, we found both topographical and non-topographical projections to subcortical structures. These data led us to propose a segregation of topographical bidirectional projections to the four fields of the pulvinar, to two subdivisions of the claustrum, and to the interlaminar portions of the LGN, structures that may operate as gates for spatial attention. The topographical efferent projections to the superficial and intermediate layers of the SC, to the thalamic reticular nucleus, and to the caudate nucleus suggest that these structures may also be involved in the processing of visual spatial attention.

Consistent with this role of the pulvinar in regulating effects of spatial attention in V4, deactivation of this portion of the pulvinar causes spatial attention deficits in monkeys (Desimone et al. 1990). Accordingly, joint electrophysiological recordings in V4 and PL reveal synchronized activity between the two structures during spatial attention tasks (Saalmann and Kastner 2011). However, several open questions remain regarding the neurophysiological basis and dynamics of pulvinar-cortical interactions. Saalmann et al. (2012) reported that neuronal oscillations of predominately lower frequencies (<30 Hz) would be responsible for synchronizing the pulvinar with cortical visual areas. However, lower-frequency oscillations have been largely associated with inactive states and sleep. Indeed, Zhou et al. (2016) used GABA to inactivate the pulvinar in monkeys performing a visual spatial task while simultaneously recording the neuronal activity of the pulvinar and V4. They observed that pulvinar inactivation drove the cortex into an "inactive" state, including reduced higher-frequency oscillations and synchronization in V4, increased lower-frequency oscillations in V4, and substantial behavioral deficits in the affected portion of the visual field. One of the established effects of attention is an enhanced synchrony in the gamma frequency band, presumably for the facilitation of neuronal communication (Fries et al. 2001; Fries 2005, 2015; Womelsdorf et al. 2007). Accordingly, Zhou et al. (2016) tested for pulvinar-V4 synchronization by measuring the spike-LFP coherence between both regions during a visual attention task. Indeed, spike-LFP coherence was found to be enhanced between the pulvinar and V4 when attention was directed to a location that fell within the recorded receptive fields. However, the authors reported that area V4 systematically led the pulvinar in terms of gamma synchrony (at least for the spatial attention task they employed). This result begs the question as to what special role (if any) the pulvinar might play in coordinating larger networks of cortical activity, as initially proposed by Shipp (2003). On the other hand, and despite the overlap in neuronal properties between V4 and the ventrolateral pulvinar, inactivation of the latter results in profound neuronal and behavioral deficits. Resolving the functional role of the pulvinar and how it interacts with the cortex in order to organize behavior will be fundamental questions to tackle in future research.

References

Adams MM, Webster MJ, Gattass R, Hof PR, Ungerleider LG (2000) Visual cortical projections and chemoarchitecture of macaque monkey pulvinar. J Comp Neurol 419:377–393

Allman JM, Kaas JH (1971) Representation of the visual field in striate and adjoining cortex of the owl monkey (*Aotus trivirgatus*). Brain Res 35:89–106

Allman JM, Kaas JH, Lane RH, Miezin FM (1972) A representation of the visual field in the inferior nucleus of the pulvinar in the owl monkey. Brain Res 40:291–302

Andersen RA, Snyder LH, Li CS, Stricanne B (1993) Coordinate transformations in the representation of spatial information. Curr Opin Neurobiol 3:171–176

Asanuma C, Andersen RA, Cowan WM (1985) The thalamic relations of the caudal inferior parietal lobule and the lateral prefrontal cortex in monkeys: divergent cortical projections from cell clusters in the medial pulvinar nucleus. J Comp Neurol 241:357–381

Baimbridge KG, Celio MR, Rogers JH (1992) Calcium-binding proteins in the nervous system. Trends Neurosci 15:303–308

Baleydier C, Morel A (1992) Segregated thalamocortical pathways to inferior parietal and inferotemporal cortex in macaque monkey. Vis Neurosci 8:391–405

Beck PD, Kaas JH (1998) Thalamic connections of the dorsomedial visual area in primates. J Comp Neurol 396:381–398

Bender DB (1981) Retinotopic organization of macaque pulvinar. J Neurophysiol 46:672–693

Bender DB, Youakim M (2001) Effect of attentive fixation in macaque thalamus and cortex. J Neurophysiol 85:219–234

Benevento LA, Davis B (1977) Topographical projections of the prestriate cortex to the pulvinar nuclei in the macaque monkey: an autoradiographic study. Exp Brain Res 30:405–424

Benevento LA, Fallon JH (1975) The ascending projections of the superior colliculus in the rhesus monkey (*Macaca mulatta*). J Comp Neurol 160:339–361

Benevento LA, Miller J (1981) Visual responses of single neurons in the caudal lateral pulvinar of the macaque monkey. J Neurosci 11:1268–1278

Benevento LA, Rezak M (1975) Extrageniculate projections to layers VI and I of striate cortex (area 17) in the rhesus monkey (*Macaca mulatta*). Brain Res 96:51–55

Benevento LA, Rezak M (1976) The cortical projections of the inferior pulvinar and adjacent lateral pulvinar in the rhesus monkey (*Macaca mulatta*): an autoradiographic study. Brain Res 108:1–24

Benevento LA, Standage GP (1983) The organization of projections of the retinorecipient and nonretinorecipient nuclei of the pretectal complex and layers of the superior colliculus to the lateral pulvinar and medial pulvinar in the macaque monkey. J Comp Neurol 217:307–336

R. Gattass et al., *The Pulvinar Thalamic Nucleus of Non-Human Primates: Architectonic and Functional Subdivisions*, Advances in Anatomy, Embryology and Cell Biology 225, https://doi.org/10.1007/978-3-319-70046-5

Benevento LA, Rezak M, Santos-Anderson R (1977) An autoradiographic study of the projections of the pretectum in the rhesus monkey (*Macaca mulatta*): evidence for sensorimotor links to the thalamus and oculomotor nuclei. Brain Res 127:197–218

Berman RA, Wurtz RH (2010) Functional identification of a pulvinar path from superior colliculus to cortical area MT. J Neurosci 30:6342–6354

Berman RA, Wurtz RH (2011) Signals conveyed in the pulvinar pathway from superior colliculus to cortical area MT. J Neurosci 31:373–384. https://doi.org/10.1523/JNEUROSCI.4738-10.2011

Bourne JA, Morrone MC (2017) Plasticity of visual pathways and function in the developing brain: is the pulvinar a crucial player? Frontier of system. Neuroscience 11:3. https://doi.org/10.3389/fnsys.2017.00003

Bridge H, Leopold DA, Bourne JA (2016) Adaptive pulvinar circuitry supports visual cognition. Trends Cogn Sci 20:146–157

Campos-Ortega JA, Hayhow WR (1972) On the organization of the visual cortical projection to the pulvinar in *Macaca mulatta*. Brain Behav Evol 6:394–423

Campos-Ortega JK, Hayhow WR, de V Clover PF (1970) A note on the problem of retinal projections to the inferior nucleus of primates. Brain Res 22:126–130

Chalupa LM, Coyle RS, Lindsley DB (1976) Effect of pulvinar lesions on visual pattern discrimination in monkeys. J Neurophysiol 39:354–369

Colby CL, Gattass R, Olson CR, Gross CG (1988) Topographical organization of cortical afferents to extrastriate visual area PO in the macaque: a dual tracer study. J Comp Neurol 269:392–413

Cowey A, Stoerig P, Bannister M (1994) Retinal ganglion cells labelled from the pulvinar nucleus in macaque monkeys. Neuroscience 61:691–705

Crick FC (1984) Function of the thalamic reticular complex: the search light hypothesis. Proc Natl Acad Sci U S A 81:4586–4590

Cusick CG, Scripter JL, Darensbourg JG, Weber JT (1993) Chemoarchitectonic subdivisions of the visual pulvinar in monkeys and their connectional relations with the middle temporal and rostral dorsolateral visual areas, MT and DLr. J Comp Neurol 336:1–30

Danziger S, Ward R, Owen V, Rafal R (2001–2002) The effects of unilateral pulvinar damage in humans on reflexive orienting and filtering of irrelevant information. Behav Neurol 13:95–104

DeFelipe J (1997) Types of neurons, synaptic connections and chemical characteristics of cells immunoreactive for calbindin-D28K, parvalbumin and calretinin in the neocortex. J Chem Neuroanat 14:1–19

Desimone R, Wessinger M, Thomas L, Schneider W (1990) Attentional control of visual perception: cortical and subcortical mechanisms. Cold Spring Harb Symp Quant Biol 55:963–971

Eidelberg E, Saldias CA (1960) A stereotaxic atlas for Cebus monkeys. J Comp Neurol 115:103–123

Federer F, Ichida JM, Jeffs J, Schiessl I, McLoughlin N, Angelucci A (2009) Four projection streams from primate V1 to the cytochrome oxidase stripes of V2. J Neurosci 29:15455–15471

Felleman DJ, Van Essen DC (1991) Distributed hierarchical processing in the primate cerebral cortex. Cereb Cortex 1:1–47

Friedmann M (1912) Die cytoarchitektonic des zwischenhirns der Cercopitheken mit bersonderer berucksichtigung des thalamus opticus. J Psychol Neurol 18:308–378

Fries P (2005) A mechanism for cognitive dynamics: neuronal communication through neuronal coherence. Trends Cogn Sci 9:474–480

Fries P (2015) Rhythms for cognition: communication through coherence. Neuron 88:220–235

Fries P, Reynolds JH, Rorie AE, Desimone R (2001) Modulation of oscillatory neuronal synchronization by selective visual attention. Science 291:1560–1563

Gattass R, Desimone R (1996) Responses of cells in the superior colliculus during performance of a spatial attention task in the macaque. Rev Bras Biol 56(Su 2):257–279

Gattass R, Desimone R (2014) Effect of microstimulation of the superior colliculus on visual space attention. J Cogn Neurosci 26:1208–1219

Gattass R, Oswaldo-Cruz E, Sousa APB (1978a) Visuotopic organization of the Cebus pulvinar: a double representation of the contralateral hemifield. Brain Res 152:1–16

Gattass R, Sousa AP, Oswaldo-Cruz E (1978b) Single unit response types in the pulvinar of the *Cebus* monkey to multisensory stimulation. Brain Res 158:75–87

Gattass R, Sousa APB, Oswaldo-Cruz E (1979) Visual receptive fields of units in the pulvinar of *Cebus* monkey. Brain Res 160:413–430

Gattass R, Nascimento-Silva S, Soares JGM, Lima B, Jansen AK, Diogo ACM, Farias MF, Marcondes M, Botelho EP, Mariani OS, Azzi J, Fiorani M (2005) Cortical visual areas in monkeys: location, topography, connections, columns, plasticity and cortical dynamics. Philos Trans R Soc Lond Ser B Biol Sci 360:709–731

Gattass R, Galkin TW, Desimone R, Ungerleider L (2014) Subcortical connections of area V4 in the macaque. J Comp Neurol 522:1941–1965

Gattass R, Lima B, Soares JGM, Ungerleider LG (2015) Controversies about the visual areas located at the anterior border of area V2 in primates. Vis Neurosci 32:E019. https://doi.org/10.1017/S0952523815000188

Glendenning KK, Hall JA, Diamond IT, Hall WC (1975) The pulvinar nucleus of *Galago senegalensis*. J Comp Neurol 161:419–458

Goldberg ME, Wurst RH (1972) Activity of superior colliculus in behaving monkey. II. Effect of attention on neuronal responses. J Neurophysiol 35:560–574

Gray D, Gutierrez C, Cusick CG (1999) Neurochemical organization of inferior pulvinar complex in squirrel monkeys and macaques revealed by acetylcholinesterase histochemistry, calbindin and CAT-301 immunostaining, and Wisteria floribunda agglutinin binding. J Comp Neurol 409:452–468

Gross CG (1991) Contribution of striate cortex and the superior colliculus to visual function in area MT, the superior temporal polysensory area and the inferior temporal cortex. Neuropsychologia 29:497–515

Gutierrez C, Cusick CG (1997) Area V1 in macaque monkeys projects to multiple histochemically defined subdivisions of the inferior pulvinar complex. Brain Res 765:349–356

Gutierrez C, Yaun A, Cusick CG (1995) Neurochemical subdivisions of the inferior pulvinar in macaque monkeys. J Comp Neurol 363:545–562

Gutierrez C, Cola MG, Seltzer B, Cusick CG (2000) Neurochemical and connectional organization of the dorsal pulvinar complex in monkeys. J Comp Neurol 419:61–86

Harting JK, Hall WC, Diamond IT (1972) Evolution of the pulvinar. Brain Behav Evol 6:424–452

Hashikawa T, Rausell E, Molinari M, Jones EG (1991) Parvalbumin- and calbindin-containing neurons in the monkey medial geniculate complex differential distribution and cortical layer specific projections. Brain Res 544:335–341

Hof PR, Morrison JH (1995) Neurofilament protein defines regional patterns of cortical organization in the macaque monkey visual system: a quantitative immunohistochemical analysis. J Comp Neurol 352:161–186

Hof PR, Glezer II, Condé F, Flagg RA, Rubin MB, Nimchinsky EA, Vogt Weisenhorn DM (1999) Cellular distribution of the calcium-binding proteins parvalbumin, calbindin, calretinin in the neocortex of mammals: phylogenetic and developmental patterns. J Chem Neuroanat 16:77–116

Holländer H (1974) Projections from the striate cortex to the diencephalon in the squirrel monkey (*Saimiri sciureus*). A light microscopic radioautographic study following intracortical injection of H^3 leucine. J Comp Neurol 155:425–440

Jones EG, Hendry SHC (1989) Differential calcium binding protein immunoreactivity distinguishes classes of relay neurons in monkey thalamic nuclei. Eur J Neurosci 1:222–246

Jones EG, Coulter JD, Hendry SH (1978) Intracortical connectivity of architectonic fields in the somatic sensory, motor and parietal cortex of monkeys. J Comp Neurol 181:291–347

Kaas JH, Lyon DC (2007) Pulvinar contributions to the dorsal and ventral streams of visual processing in primates. Brain Res Rev 55:285–296

LaBerge D, Buchsbaum MS (1990) Positron emission tomography measurements of pulvinar activity during an attention task. J Neurosci 10:613–619
Levitt JB, Yoshioka T, Lund JS (1995) Connections between the pulvinar complex and cytochrome oxidase-defined compartments in visual area V2 of macaque monkey. Exp Brain Res 104:419–430
Lima B, Singer W, Neuenschwander S (2011) Gamma responses correlate with temporal expectation in monkey primary visual cortex. J Neurosci 31:15919–15931
Lin CS, Kaas JH (1979) The inferior pulvinar complex in owl monkeys: architectonic subdivisions and patterns of input from the superior colliculus and subdivisions of visual cortex. J Comp Neurol 187:655–678
Lin CS, Wagor E, Kaas JH (1974) Projections from the pulvinar to the middle temporal visual area (MT) in the owl monkey, Aotus trivirgatus. Brain Res 76:145–149
Lund JS, Boothe RG (1975) Interlaminar connections and pyramidal neuron organization in the visual cortex, area 17 of the Macaque monkey. J Comp Neurol 159:305–344
Lyon DC, Nassi JJ, Callaway EM (2010) A disynaptic relay from superior colliculus to dorsal stream visual cortex in macaque monkey. Neuron 65:270–279
Lysakowski A, Standage GP, Benevento LA (1986) Histochemical and architectonic differentiation of zones of pretectal and collicular inputs to the pulvinar and dorsal lateral geniculate nuclei in the macaque. J Comp Neurol 250:431–448
Marion R, Li K, Purushothaman G, Jiang Y, Casagrande VA (2013) Morphological and neurochemical comparisons between pulvinar and V1 projections to V2. J Comp Neurol 521:813–832
Mathers LH (1971) Tectal projection to the posterior thalamus of the squirrel monkey. Brain Res 35:295–298
Mathers LH (1972) Ultrastructure of the pulvinar of the squirrel monkey. J Comp Neurol 146:15–42
Mathers LH, Rapisardi SC (1973) Visual and somatosensory receptive fields of neurons in the squirrel monkey pulvinar. Brain Res 64:65–83
Mishkin M, Ungerleider LG (1982) Contribution of striate inputs to the visuospatial functions of parieto-preoccipital cortex in monkeys. Behav Brain Res 6:57–77
Nakamura RK, Mishkin M (1986) Chronic 'blindness' following lesions of nonvisual cortex in the monkey. Exp Brain Res 63:173–184
O'Brien BJ, Abel PL, Olavarria JF (2001) The retinal input to calbindin-D28k-defined subdivisions in macaque inferior pulvinar. Neurosci Lett 312:145–148
Ogren MP (1977) Evidence for a projection from pulvinar to striate cortex in the squirrel monkey (*Saimiri sciureus*). Exp Neurol 54:622–625
Ogren MP, Hendrickson AE (1975) Afferent and efferent pathways of striate cortex in squirrel and rhesus monkey. Anat Rec 181:439
Ogren MP, Hendrickson AE (1976) Pathways between striate cortex and subcortical regions in *Macaca mulatta* and *Saimiri sciureus*: evidence for a reciprocal pulvinar connection. Exp Neurol 53:780–800
Ogren MP, Hendrickson AE (1977) The distribution of pulvinar terminals in visual areas 17 and 18 of the monkey. Brain Res 137:343–350
Ogren MP, Hendrickson AE (1979) The structural organization of the inferior and lateral subdivisions of the Macaca monkey pulvinar. J Comp Neurol 188:147–178
Olshausen BA, Anderson CH, Van Essen DC (1993) A neurobiological model of visual attention and invariant pattern recognition based on dynamic routing of information. J Neurosci 13:4700–4719
Olszewski J (1952) The thalamus of the *Macaca mulatta* – an Atlas for use with the stereotaxic instrument. S. Karger, Basel, 93 p
Partlow GD, Colonnier M, Szabo J (1977) Thalamic projections of the superior colliculus in the rhesus monkey, *Macaca mulatta*: a light and electron microscopic study. J Comp Neurol 171:285–318

Petersen SE, Robinson DL, Keys W (1985) Pulvinar nuclei of the behaving rhesus monkey: visual response and their modulation. J Neurophysiol 54:867–885
Petersen SE, Robinson DL, Morris JD (1987) Contributions of the pulvinar to visual spatial attention. Neuropsychologia 25:97–105
Posner MI, Petersen SE (1990) The attention system of the human brain. Annu Rev Neurosci 13:25–42
Purushothaman G, Marion R, Li K, Casagrande VA (2012) Gating and control of primary visual cortex by pulvinar. Nat Neurosci 15:905–912
Rafal RD, Posner MI (1987) Deficits in human visual spatial attention following thalamic lesions. Proc Natl Acad Sci U S A 84:7349–7353
Rakic P (1974) Embryonic development of the pulvinar – LP complex in man. In: Cooper IS, Riklan M, Rakic P (eds) The pulvinar – LP complex. Charles C. Thomas, Springfield, IL, pp 3–35
Rezak M, Benevento LA (1979) A comparison of the organization of the projections of the dorsal lateral geniculate nucleus, the inferior pulvinar and the adjacent lateral pulvinar to primary visual cortex (area 17) in the macaque monkey. Brain Res 167:19–40
Robinson DL, Petersen SE (1992) The pulvinar and visual salience. Trends Neurosci 15:127–132
Rockland KS, Pandya DN (1979) Laminar origins and terminations of cortical connections of the occipital lobe in the rhesus monkey. Brain Res 179:3–20
Saalmann YB, Kastner S (2011) Cognitive and perceptual functions of the visual thalamus. Neuron 71:209–223
Saalmann YB, Pinsk MA, Wang L, Li X, Kastner S (2012) The pulvinar regulates information transmission between cortical areas based on attention demands. Science 337(6095):753–756
Sherman SM, Guillery RW (2002) The role of the thalamus in the flow of information to the cortex. Philos Trans R Soc Lond Ser B Biol Sci 357:1695–1708
Shipp S (2000) A new anatomical basis for 'spotlight' metaphors of attention. Eur J Neurosci 12 (Suppl 11):196
Shipp S (2003) The functional logic of cortico-pulvinar connections. Philos Trans R Soc Lond Ser B Biol Sci 358:1605–1624
Sincich LC, Horton JC (2002) Pale cytochrome oxidase stripes in V2 receive the richest projection from macaque striate cortex. J Comp Neurol 447:18–33
Siqueira EB (1971) The cortical connections of the nucleus pulvinaris of the dorsal thalamus in the rhesus monkey. Int J Neurol 8:139–154
Soares JGM, Gattass R, Souza APB, Rosa MGP, Fiorani M Jr, Brandão BL (2001) Connectional and neurochemical subdivisions of the pulvinar in Cebus monkeys. Vis Neurosci 18:25–41
Soares JGM, Diogo ACM, Fiorani M, Souza APB, Gattass R (2004) Effects of inactivation of the lateral pulvinar on response properties of second visual area cells in Cebus monkeys. Clin Exp Pharmacol Physiol 31:580–590
Spatz WB, Erdmann G (1974) Striate cortex projections to the lateral geniculate and other thalamic nuclei; a study using degeneration and autoradiographic tracing methods in the marmoset Callithrix. Brain Res 82:91–108
Standage GP, Benevento LA (1983) The organization of connections between the pulvinar and visual area MT in the macaque monkey. Brain Res 262:288–294
Steele GE, Weller RE (1993) Subcortical connections of subdivisions of inferior temporal cortex in squirrel monkeys. Vis Neurosci 10:563–583
Stepniewska I, Kaas JH (1997) Architectonic subdivisions of the inferior pulvinar in NewWorld and OldWorld monkeys. Vis Neurosci 14:1043–1060
Sternberger LA, Sternberger NH (1983) Monoclonal antibodies distinguish phosphorylated and nonphosphorylated forms of neurofilaments in situ. Neurobiology 80:6126–6130
Trageser JC, Keller A (2004) Reducing the uncertainty: gating of peripheral inputs by zona incerta. J Neurosci 24:8911–8915
Trageser JC, Burke KA, Masri R, Li Y, Sellers L, Keller A (2006) State-dependent gating of sensory inputs by zona incerta. J Neurophysiol 96:1456–1463

Treue S, Maunsell JH (1996) Attentional modulation of visual motion processing in cortical areas MT and MST. Nature 382:539–541

Trojanowski JQ, Jacobson S (1974) Medial pulvinar afferents to frontal eye fields in rhesus monkey demonstrated by horseradish peroxidase. Brain Res 80:395–411

Trojanowski JQ, Jacobson S (1975) Peroxidase labeled subcortical pulvinar afferents in rhesus monkey. Brain Res 97:144–150

Trojanowski JQ, Jacobson S (1976) Areal and laminar distribution of some pulvinar cortical efferents in rhesus monkey. J Comp Neurol 169:371–392

Ungerleider LG, Christensen CA (1977) Pulvinar lesions in monkeys produce abnormal eye movements during visual discrimination training. Brain Res 136:189–196

Ungerleider LG, Galkin TW, Mishkin M (1983) Visuotopic organization of projections of striate cortex to inferior and lateral pulvinar in rhesus monkey. J Comp Neurol 217:137–157

Ungerleider LG, Desimone R, Galkin TW, Mishkin M (1984) Subcortical projections of area MT in the macaque. J Comp Neurol 223:368–386

Ungerleider LG, Galkin TW, Desimone R, Gattass R (2008) Cortical connections of area V4 in the macaque. Cereb Cortex 18:477–499

Ungerleider LG, Galkin TW, Desimone R, Gattass R (2014) Subcortical projections of area V2 in the macaque. J Cogn Neurosci 26:1220–1233

Walker AE (1938) The primate thalamus. University of Chicago Press, Chicago, IL

Ward R, Danziger S, Owen V, Rafal R (2002) Deficits in spatial coding and feature binding following damage to spatiotopic maps in the human pulvinar. Nat Neurosci 5:99–100

Warner CE, Goldshmit Y, Bourne JA (2010) Retinal afferents synapse with relay cells targeting the middle temporal area in the pulvinar and lateral geniculate nuclei. Front Neuroanat 4:8. https://doi.org/10.3389/neuro.05.008.2010

Womelsdorf T, Schoffelen J-M, Oostenveld R, Singer W, Desimone R, Engel AK, Fries P (2007) Modulation of neuronal interactions through neuronal synchronization. Science 316:1609–1612

Wong-Riley MTT (1977) Connections between the pulvinar nucleus and the prestriate cortex in the squirrel monkey as revealed by peroxidase histochemistry and autoradiography. Brain Res 134:249–267

Zeki S, Shipp S (1989) Modular connections between areas V2 and V4 of macaque monkey visual cortex. Eur J Neurosci 1:494–506

Zhou H, Schafer RJ, Desimone R (2016) Pulvinar-cortex interactions in vision and attention. Neuron 89:209–220

Kolibri 360 GmbH
Ferdinand-Jühlke-Straße 7
99095 Erfurt, Deutschland
produktsicherheit@kolibri360.de